Learn, Practice, Succeed

Eureka Math®

Grade 7
Module 4

Published by Great Minds®

Printed in the U.S.A.

This book may be purchased from the publisher at eureka-math.org.
10 9 8 7 6 5 4 3 2 1

ISBN 978-1-64054-975-3

G7-M4-LPS-05.2019

Students, families, and educators:

Thank you for being part of the *Eureka Math®* community, where we celebrate the joy, wonder, and thrill of mathematics.

In *Eureka Math* classrooms, learning is activated through rich experiences and dialogue. That new knowledge is best retained when it is reinforced with intentional practice. The *Learn, Practice, Succeed* book puts in students' hands the problem sets and fluency exercises they need to express and consolidate their classroom learning and master grade-level mathematics. Once students learn and practice, they know they can succeed.

What is in the Learn, Practice, Succeed *book?*

Fluency Practice: Our printed fluency activities utilize the format we call a Sprint. Instead of rote recall, Sprints use patterns across a sequence of problems to engage students in reasoning and to reinforce number sense while building speed and accuracy. Sprints are inherently differentiated, with problems building from simple to complex. The tempo of the Sprint provides a low-stakes adrenaline boost that increases memory and automaticity.

Classwork: A carefully sequenced set of examples, exercises, and reflection questions support students' in-class experiences and dialogue. Having classwork preprinted makes efficient use of class time and provides a written record that students can refer to later.

Exit Tickets: Students show teachers what they know through their work on the daily Exit Ticket. This check for understanding provides teachers with valuable real-time evidence of the efficacy of that day's instruction, giving critical insight into where to focus next.

Homework Helpers and Problem Sets: The daily Problem Set gives students additional and varied practice and can be used as differentiated practice or homework. A set of worked examples, Homework Helpers, support students' work on the Problem Set by illustrating the modeling and reasoning the curriculum uses to build understanding of the concepts the lesson addresses.

Homework Helpers and Problem Sets from prior grades or modules can be leveraged to build foundational skills. When coupled with *Affirm®*, *Eureka Math*'s digital assessment system, these Problem Sets enable educators to give targeted practice and to assess student progress. Alignment with the mathematical models and language used across *Eureka Math* ensures that students notice the connections and relevance to their daily instruction, whether they are working on foundational skills or getting extra practice on the current topic.

Where can I learn more about Eureka Math *resources?*

The Great Minds® team is committed to supporting students, families, and educators with an ever-growing library of resources, available at eureka-math.org. The website also offers inspiring stories of success in the *Eureka Math* community. Share your insights and accomplishments with fellow users by becoming a *Eureka Math* Champion.

Best wishes for a year filled with "aha" moments!

Jill Diniz

Jill Diniz
Chief Academic Officer, Mathematics
Great Minds

Contents

Module 4: Percent and Proportional Relationships

Number Correct: ______

Fractions, Decimals, and Percents—Round 1

Directions: Write each number in the alternate form indicated.

1.	$\frac{20}{100}$ as a percent	
2.	$\frac{40}{100}$ as a percent	
3.	$\frac{80}{100}$ as a percent	
4.	$\frac{85}{100}$ as a percent	
5.	$\frac{95}{100}$ as a percent	
6.	$\frac{100}{100}$ as a percent	
7.	$\frac{10}{10}$ as a percent	
8.	$\frac{1}{1}$ as a percent	
9.	$\frac{1}{10}$ as a percent	
10.	$\frac{2}{10}$ as a percent	
11.	$\frac{4}{10}$ as a percent	
12.	75% as a decimal	
13.	25% as a decimal	
14.	15% as a decimal	
15.	10% as a decimal	
16.	5% as a decimal	
17.	30% as a fraction	
18.	60% as a fraction	
19.	90% as a fraction	
20.	50% as a fraction	
21.	25% as a fraction	
22.	20% as a fraction	

23.	$\frac{9}{10}$ as a percent	
24.	$\frac{9}{20}$ as a percent	
25.	$\frac{9}{25}$ as a percent	
26.	$\frac{9}{50}$ as a percent	
27.	$\frac{9}{75}$ as a percent	
28.	$\frac{18}{75}$ as a percent	
29.	$\frac{36}{75}$ as a percent	
30.	96% as a fraction	
31.	92% as a fraction	
32.	88% as a fraction	
33.	44% as a fraction	
34.	22% as a fraction	
35.	3% as a decimal	
36.	30% as a decimal	
37.	33% as a decimal	
38.	33.3% as a decimal	
39.	3.3% as a decimal	
40.	0.3% as a decimal	
41.	$\frac{1}{3}$ as a percent	
42.	$\frac{1}{9}$ as a percent	
43.	$\frac{2}{9}$ as a percent	
44.	$\frac{8}{9}$ as a percent	

Number Correct: ______
Improvement: ______

Fractions, Decimals, and Percents—Round 2

Directions: Write each number in the alternate form indicated.

1.	$\frac{30}{100}$ as a percent		23.	$\frac{6}{10}$ as a percent	
2.	$\frac{60}{100}$ as a percent		24.	$\frac{6}{20}$ as a percent	
3.	$\frac{70}{100}$ as a percent		25.	$\frac{6}{25}$ as a percent	
4.	$\frac{75}{100}$ as a percent		26.	$\frac{6}{50}$ as a percent	
5.	$\frac{90}{100}$ as a percent		27.	$\frac{6}{75}$ as a percent	
6.	$\frac{50}{100}$ as a percent		28.	$\frac{12}{75}$ as a percent	
7.	$\frac{5}{10}$ as a percent		29.	$\frac{24}{75}$ as a percent	
8.	$\frac{1}{2}$ as a percent		30.	64% as a fraction	
9.	$\frac{1}{4}$ as a percent		31.	60% as a fraction	
10.	$\frac{1}{8}$ as a percent		32.	56% as a fraction	
11.	$\frac{3}{8}$ as a percent		33.	28% as a fraction	
12.	60% as a decimal		34.	14% as a fraction	
13.	45% as a decimal		35.	9% as a decimal	
14.	30% as a decimal		36.	90% as a decimal	
15.	6% as a decimal		37.	99% as a decimal	
16.	3% as a decimal		38.	99.9% as a decimal	
17.	3% as a fraction		39.	9.9% as a decimal	
18.	6% as a fraction		40.	0.9% as a decimal	
19.	60% as a fraction		41.	$\frac{4}{9}$ as a percent	
20.	30% as a fraction		42.	$\frac{5}{9}$ as a percent	
21.	45% as a fraction		43.	$\frac{2}{3}$ as a percent	
22.	15% as a fraction		44.	$\frac{1}{6}$ as a percent	

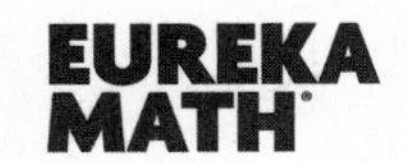

Opening Exercise 1: Matching

Match the percents with the correct sentence clues.

25%	I am half of a half. 5 cubic inches of water filled in a 20 cubic inch bottle.
50%	I am less than $\frac{1}{100}$. 25 out of $5{,}000$ contestants won a prize.
30%	I am the chance of birthing a boy or a girl. Flip a coin, and it will land on heads or tails.
1%	I am less than a half but more than one-fourth. 15 out of 50 play drums in a band.
10%	I am equal to 1. 35 question out of 35 questions were answered correctly.
100%	I am more than 1. Instead of the $\$1{,}200$ expected to be raised, $\$3{,}600$ was collected for the school's fundraiser.
300%	I am a tenth of a tenth. One penny is this part of one dollar.
$\frac{1}{2}\%$	I am less than a fourth but more than a hundredth. $\$11$ out of $\$110$ earned is saved in the bank.

Opening Exercise 2

Color in the grids to represent the following fractions:

a. $\frac{30}{100}$

b. $\frac{3}{100}$

c. $\frac{\frac{1}{3}}{100}$

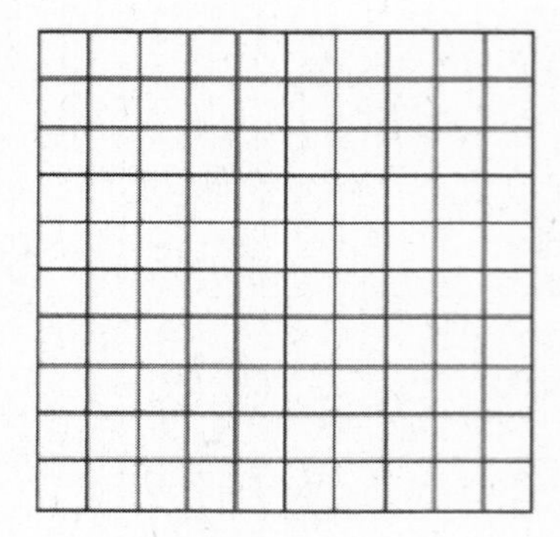

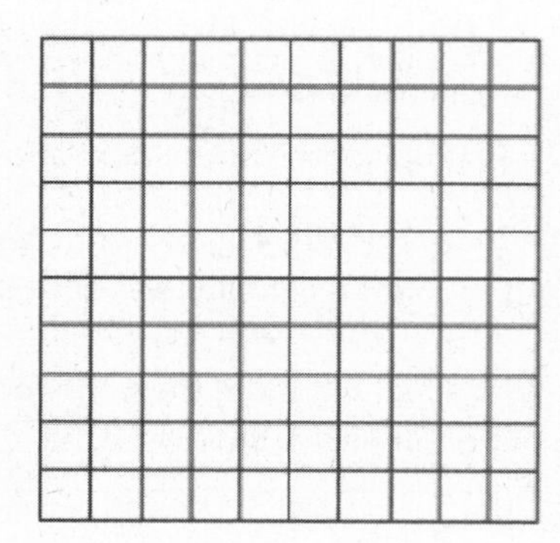

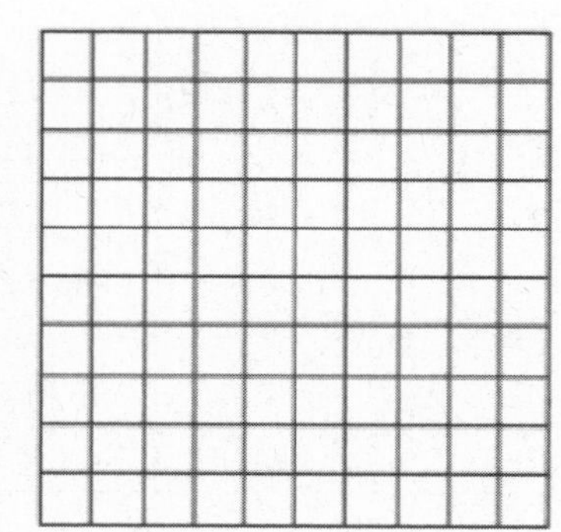

Example 1

Use the definition of the word *percent* to write each percent as a fraction and then as a decimal.

Percent	Fraction	Decimal
37.5%		
100%		
110%		
1%		
$\frac{1}{2}\%$		

Example 2

Fill in the chart by converting between fractions, decimals, and percents. Show your work in the space below.

Fraction	Decimal	Percent
		350%
	0.025	
$\frac{1}{8}$		

Exercise: Class Card Activity

Read your card to yourself (each student has a different card), and work out the problem. When the exercise begins, listen carefully to the questions being read. When you have the card with the equivalent value, respond by reading your card aloud.

Examples:

0.22 should be read "twenty-two hundredths."

$\frac{\frac{1}{5}}{1000}$ should be read "one-fifth thousandths" or "one-fifth over one thousand."

$\frac{7}{300}$ should be read "seven three-hundredths" or "seven over three hundred."

$\frac{200}{100}$ should be read "two hundred hundredths" or "two hundred over one hundred."

Lesson Summary

- One percent is the number $\frac{1}{100}$ and is written 1%. The number $P\%$ is the same as the number $\frac{P}{100}$.
- Usually, there are three ways to write a number: a percent, a fraction, and a decimal. The fraction and decimal forms of $P\%$ are equivalent to $\frac{P}{100}$.

Name ____ Date ____________________

1. Fill in the chart converting between fractions, decimals, and percents. Show work in the space provided.

Fraction	Decimal	Percent
$\frac{1}{8}$		
	1.125	
		$\frac{2}{5}\%$

2. Using the values from the chart in Problem 1, which is the least and which is the greatest? Explain how you arrived at your answers.

1. Create a model to represent the following percents.

   a. 75%

Each box represents 1% because there are 100 boxes. Therefore, I can shade in any 75 boxes to represent 75%.

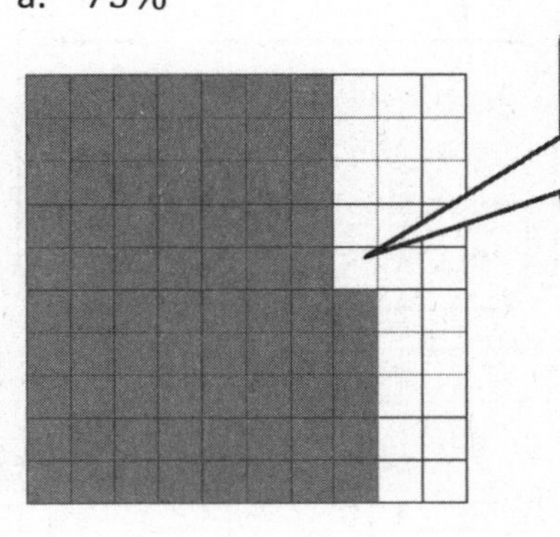

   b. 0.5%

I will have to shade in less than one box because the percent given is less than 1.

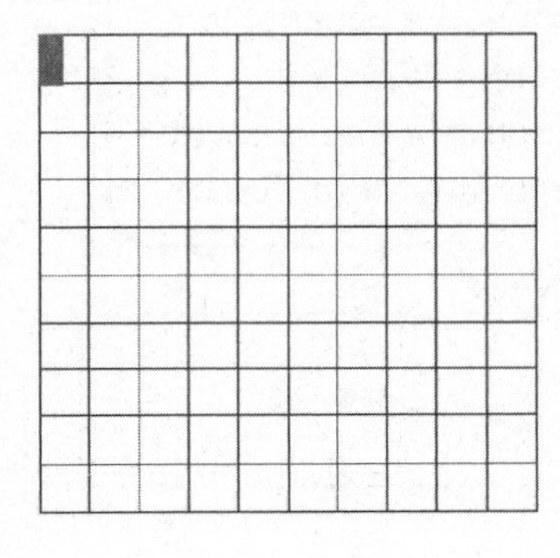

   c. 350%

350% is greater than 100%, so I will need more than one grid to model the given percent.

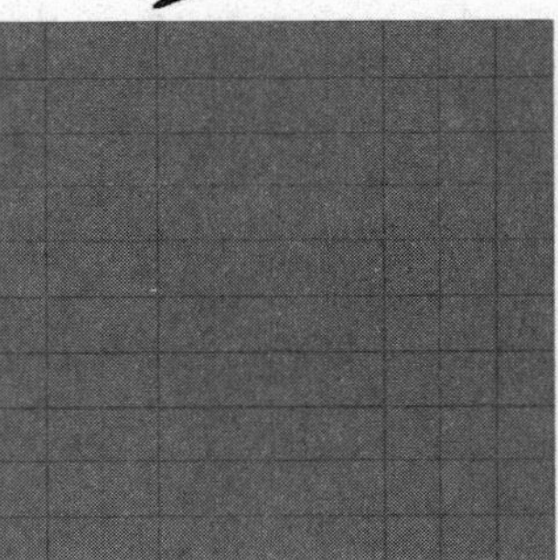

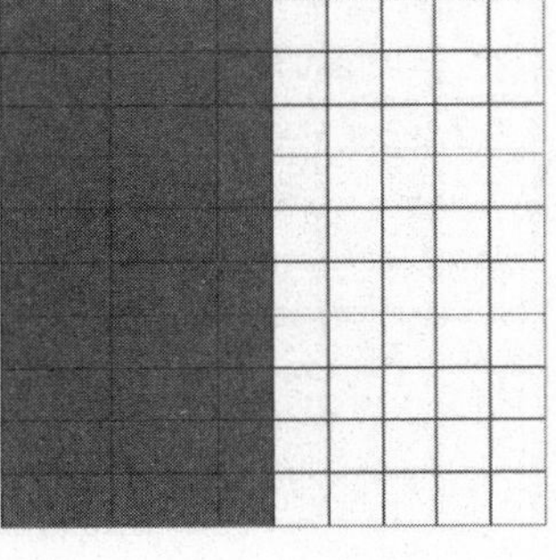

2. Complete the table by converting among fractions, decimals, and percents.

To convert a fraction to a decimal, I can either use long division or find an equivalent fraction with a denominator as a multiple of 10.

To convert a fraction to a percent, I find an equivalent fraction with a denominator of 100.

Fraction	Decimal	Percent
$\frac{1}{5}$	$\frac{1}{5}=\frac{2}{10}=0.2$	$\frac{1}{5}=\frac{20}{100}=20\%$

To convert a decimal to a fraction, I use the place value of the digit furthest to the right to determine my denominator.

The 5 is in the thousandths place.

To convert a decimal to a percent, I write the decimal as a fraction with a denominator of 100.

$\frac{815}{1000}$	0.815	$\frac{81.5}{100}=81.5\%$

To convert a percent to a fraction, I write the percent as a fraction with a denominator of 100 and change to a mixed number if necessary.

To convert a percent to a decimal, I write the percent as a fraction with a denominator of 100 and then use place value to write the value as a decimal.

$\frac{225}{100}=2\frac{25}{100}=2\frac{1}{4}$	$\frac{225}{100}=2\frac{25}{100}=2.25$	225%

3. Order the following from least to greatest.

$200\%,\ 2.1,\ \frac{1}{50},\ 0.2,\ \frac{20{,}000}{1{,}000},\ 0.02\%,\ 0.002$

I can rewrite every term as a decimal to make the comparison easier:
$2,\ 2.1,\ 0.02,\ 0.2,\ 20,\ 0.0002,\ 0.002$

$\mathbf{0.02\%,\ 0.002,\ \frac{1}{50},\ 0.2,\ 200\%,\ 2.1,\ \frac{20{,}000}{1{,}000}}$

1. Create a model to represent the following percents.

   a. 90% b. 0.9% c. 900% d. $\frac{9}{10}\%$

2. Benjamin believes that $\frac{1}{2}\%$ is equivalent to 50%. Is he correct? Why or why not?

3. Order the following from least to greatest:

   $100\%, \frac{1}{100}, 0.001\%, \frac{1}{10}, 0.001, 1.1, 10$, and $\frac{10{,}000}{100}$

4. Fill in the chart by converting between fractions, decimals, and percents. Show work in the space below.

Fraction	Decimal	Percent
		100%
	0.0825	
	6.25	
		$\frac{1}{8}\%$
$\frac{2}{300}$		
		33.3%
$\frac{\frac{3}{4}}{100}$		
		250%
	0.005	
$\frac{150}{100}$		
	0.055	

Opening Exercise

a. What is the whole unit in each scenario?

Scenario	Whole Unit
15 is what percent of 90?	
What number is 10% of 56?	
90% of a number is 180.	
A bag of candy contains 300 pieces and 25% of the pieces in the bag are red.	
Seventy percent (70%) of the students earned a B on the test.	
The 20 girls in the class represented 55% of the students in the class.	

b. Read each problem, and complete the table to record what you know.

Problem	Part	Percent	Whole
40% of the students on the field trip love the museum. If there are 20 students on the field trip, how many love the museum?			
40% of the students on the field trip love the museum. If 20 students love the museum, how many are on the field trip?			
20 students on the field trip love the museum. If there are 40 students on the field trip, what percent love the museum?			

Example 1: Visual Approaches to Finding a Part, Given a Percent of the Whole

In Ty's math class, 20% of students earned an A on a test. If there were 30 students in the class, how many got an A?

Exercise 1

In Ty's art class, 12% of the Flag Day art projects received a perfect score. There were 25 art projects turned in by Ty's class. How many of the art projects earned a perfect score? (Identify the whole.)

Example 2: A Numeric Approach to Finding a Part, Given a Percent of the Whole

In Ty's English class, 70% of the students completed an essay by the due date. There are 30 students in Ty's English class. How many completed the essay by the due date?

Example 3: An Algebraic Approach to Finding a Part, Given a Percent of the Whole

A bag of candy contains 300 pieces of which 28% are red. How many pieces are red?

Which quantity represents the whole?

Which of the terms in the percent equation is unknown? Define a letter (variable) to represent the unknown quantity.

Write an expression using the percent and the whole to represent the number of pieces of red candy.

Write and solve an equation to find the unknown quantity.

Exercise 2

A bag of candy contains 300 pieces of which 28% are red. How many pieces are not red?

a. Write an equation to represent the number of pieces that are not red, n.

b. Use your equation to find the number of pieces of candy that are not red.

c. Jah-Lil told his math teacher that he could use the answer from Example 3 and mental math to find the number of pieces of candy that are not red. Explain what Jah-Lil meant by that.

Example 4: Comparing Part of a Whole to the Whole with the Percent Formula

Zoey inflated 24 balloons for decorations at the middle school dance. If Zoey inflated 15% of the total number of balloons inflated for the dance, how many balloons are there total? Solve the problem using the percent formula, and verify your answer using a visual model.

Example 5: Finding the Percent Given a Part of the Whole and the Whole

Haley is making admission tickets to the middle school dance. So far she has made 112 tickets, and her plan is to make 320 tickets. What percent of the admission tickets has Haley produced so far? Solve the problem using the percent formula, and verify your answer using a visual model.

Lesson Summary

- Visual models or numeric methods can be used to solve percent problems.
- An equation can be used to solve percent problems:

$$\text{Part} = \text{Percent} \times \text{Whole}.$$

Name ______________________________ Date ______________

1. On a recent survey, 60% of those surveyed indicated that they preferred walking to running.

   a. If 540 people preferred walking, how many people were surveyed?

   b. How many people preferred running?

2. Which is greater: 25% of 15 or 15% of 25? Explain your reasoning using algebraic representations or visual models.

Represent each situation using an equation.

1. What number is 20% of 80?

> I know that "is" means "equals" and that "of" means "multiply." Using this knowledge, I can translate the words into an equation.

Let n represent the unknown number.

$n = 20\%(80)$
$n = 0.2(80)$
$n = 16$

> Before I can solve for n, I need to convert the percent to a decimal or a fraction.

This means $\mathbf{16}$ ***is*** $\mathbf{20\%}$ ***of*** $\mathbf{80}$.

2. 28 is 40% of what number?

Let n represent the unknown number.

$$\begin{aligned} 28 &= 40\%(n) \\ 28 &= \frac{40}{100}n \\ \left(\frac{100}{40}\right)(28) &= \left(\frac{100}{40}\right)\left(\frac{40}{100}n\right) \\ 70 &= n \end{aligned}$$

Therefore, $\mathbf{28}$ ***is*** $\mathbf{40\%}$ ***of*** $\mathbf{70}$.

> In order to solve for n, I multiply both sides of the equation by the multiplicative inverse of $\frac{40}{100}$.

3. 40 is what percent of 50?

Let p represent the unknown percent.

$$\begin{aligned} 40 &= p(50) \\ 40\left(\frac{1}{50}\right) &= p(50)\left(\frac{1}{50}\right) \\ 0.8 &= p \end{aligned}$$

$p = 0.8 = 80\%$

Therefore, $\mathbf{40}$ ***is*** $\mathbf{80\%}$ ***of*** $\mathbf{50}$.

Use percents to solve the following real-world problems.

> When solving real-world problems, I can use the formula Part $=$ Percent $\times$ Whole.

4. Michael spent 40% of his money on new shoes. If Michael spent \$85 on his new shoes, how much money did Michael have at the beginning?

Part = Percent × Whole

> I know the percent and the part; I need to calculate the whole.

Let w represent the amount of money Michael had at the beginning.

$$85 = 40\% \times w$$

> Michael spent 40% of his money, w, on shoes.

$$85 = 0.4 \times w$$

$$85\left(\frac{1}{0.4}\right) = 0.4w\left(\frac{1}{0.4}\right)$$

$$212.5 = w$$

This means that Michael had $\$212.50$ before he bought new shoes.

5. McKayla took 30 shots during her last basketball game, but only made 18 baskets. What percent of her shots did McKayla make?

Part = Percent × Whole

> I know the whole is the number of shots McKayla took, and the part is the 18 baskets she made.

Let p represent the percent of the shots McKayla made.

$$18 = p \times 30$$

$$18\left(\frac{1}{30}\right) = p(30)\left(\frac{1}{30}\right)$$

$$0.6 = p$$

$p = 0.6 = 60\%$

This means that McKayla made 60% of the baskets she shot.

1. Represent each situation using an equation. Check your answer with a visual model or numeric method.
   a. What number is 40% of 90?
   b. What number is 45% of 90?
   c. 27 is 30% of what number?
   d. 18 is 30% of what number?
   e. 25.5 is what percent of 85?
   f. 21 is what percent of 60?

2. 40% of the students on a field trip love the museum. If there are 20 students on the field trip, how many love the museum?

3. Maya spent 40% of her savings to pay for a bicycle that cost her $\$85$.
   a. How much money was in her savings to begin with?
   b. How much money does she have left in her savings after buying the bicycle?

4. Curtis threw 15 darts at a dartboard. 40% of his darts hit the bull's-eye. How many darts did not hit the bull's-eye?

5. A tool set is on sale for $\$424.15$. The original price of the tool set was $\$499.00$. What percent of the original price is the sale price?

6. Matthew scored a total of 168 points in basketball this season. He scored 147 of those points in the regular season, and the rest were scored in his only playoff game. What percent of his total points did he score in the playoff game?

7. Brad put 10 crickets in his pet lizard's cage. After one day, Brad's lizard had eaten 20% of the crickets he had put in the cage. By the end of the next day, the lizard had eaten 25% of the remaining crickets. How many crickets were left in the cage at the end of the second day?

8. A furnace used 40% of the fuel in its tank in the month of March and then used 25% of the remaining fuel in the month of April. At the beginning of March, there were 240 gallons of fuel in the tank. How much fuel (in gallons) was left at the end of April?

9. In Lewis County, there were 2,277 student athletes competing in spring sports in 2014. That was 110% of the number from 2013, which was 90% of the number from the year before. How many student athletes signed up for a spring sport in 2012?

10. Write a real-world word problem that could be modeled by the equation below. Identify the elements of the percent equation and where they appear in your word problem, and then solve the problem.

$$57.5 = p(250)$$

Number Correct: ______

Part, Whole, or Percent—Round 1

Directions: Find each missing value.

1.	1% of 100 is?	
2.	2% of 100 is?	
3.	3% of 100 is?	
4.	4% of 100 is?	
5.	5% of 100 is?	
6.	9% of 100 is?	
7.	10% of 100 is?	
8.	10% of 200 is?	
9.	10% of 300 is?	
10.	10% of 500 is?	
11.	10% of 550 is?	
12.	10% of 570 is?	
13.	10% of 470 is?	
14.	10% of 170 is?	
15.	10% of 70 is?	
16.	10% of 40 is?	
17.	10% of 20 is?	
18.	10% of 25 is?	
19.	10% of 35 is?	
20.	10% of 36 is?	
21.	10% of 37 is?	
22.	10% of 37.5 is?	

23.	10% of 22 is?	
24.	20% of 22 is?	
25.	30% of 22 is?	
26.	50% of 22 is?	
27.	25% of 22 is?	
28.	75% of 22 is?	
29.	80% of 22 is?	
30.	85% of 22 is?	
31.	90% of 22 is?	
32.	95% of 22 is?	
33.	5% of 22 is?	
34.	15% of 80 is?	
35.	15% of 60 is?	
36.	15% of 40 is?	
37.	30% of 40 is?	
38.	30% of 70 is?	
39.	30% of 60 is?	
40.	45% of 80 is?	
41.	45% of 120 is?	
42.	120% of 40 is?	
43.	120% of 50 is?	
44.	120% of 55 is?	

Number Correct: ______
Improvement: ______

Part, Whole, or Percent—Round 2

Directions: Find each missing value.

1.	20% of 100 is?	
2.	21% of 100 is?	
3.	22% of 100 is?	
4.	23% of 100 is?	
5.	25% of 100 is?	
6.	25% of 200 is?	
7.	25% of 300 is?	
8.	25% of 400 is?	
9.	25% of 4,000 is?	
10.	50% of 4,000 is?	
11.	10% of 4,000 is?	
12.	10% of 4,700 is?	
13.	10% of 4,600 is?	
14.	10% of 4,630 is?	
15.	10% of 463 is?	
16.	10% of 46.3 is?	
17.	10% of 18 is?	
18.	10% of 24 is?	
19.	10% of 3.63 is?	
20.	10% of 0.363 is?	
21.	10% of 37 is?	
22.	10% of 37.5 is?	

23.	10% of 4 is?	
24.	20% of 4 is?	
25.	30% of 4 is?	
26.	50% of 4 is?	
27.	25% of 4 is?	
28.	75% of 4 is?	
29.	80% of 4 is?	
30.	85% of 4 is?	
31.	90% of 4 is?	
32.	95% of 4 is?	
33.	5% of 4 is?	
34.	15% of 40 is?	
35.	15% of 30 is?	
36.	15% of 20 is?	
37.	30% of 20 is?	
38.	30% of 50 is?	
39.	30% of 90 is?	
40.	45% of 90 is?	
41.	90% of 120 is?	
42.	125% of 40 is?	
43.	125% of 50 is?	
44.	120% of 60 is?	

Opening Exercise

If each 10×10 unit square represents one whole, then what percent is represented by the shaded region?

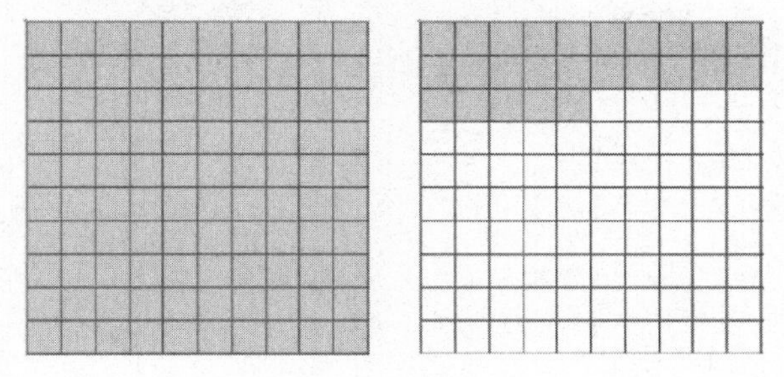

In the model above, 25% represents a quantity of 10 students. How many students does the shaded region represent?

Example

a. The members of a club are making friendship bracelets to sell to raise money. Anna and Emily made 54 bracelets over the weekend. They need to produce 300 bracelets by the end of the week. What percent of the bracelets were they able to produce over the weekend?

b. Anna produced 32 bracelets of the 54 bracelets produced by Emily and Anna over the weekend. Write the number of bracelets that Emily produced as a percent of those that Anna produced.

c. Write the number of bracelets that Anna produced as a percent of those that Emily produced.

Exercises

1. There are 750 students in the seventh-grade class and 625 students in the eighth-grade class at Kent Middle School.

   a. What percent is the seventh-grade class of the eighth-grade class at Kent Middle School?

   b. The principal will have to increase the number of eighth-grade teachers next year if the seventh-grade enrollment exceeds 110% of the current eighth-grade enrollment. Will she need to increase the number of teachers? Explain your reasoning.

2. At Kent Middle School, there are 104 students in the band and 80 students in the choir. What percent of the number of students in the choir is the number of students in the band?

3. At Kent Middle School, breakfast costs $\$1.25$ and lunch costs $\$3.75$. What percent of the cost of lunch is the cost of breakfast?

4. Describe a real-world situation that could be modeled using the equation $398.4 = 0.83(x)$. Describe how the elements of the equation correspond with the real-world quantities in your problem. Then, solve your problem.

Lesson Summary

- Visual models or arithmetic methods can be used to solve problems that compare quantities with percents.
- Equations can be used to solve percent problems using the basic equation

$$\text{Quantity} = \text{Percent} \times \text{Whole}.$$

- *Quantity* in the new percent formula is the equivalent of *part* in the original percent formula.

Name ______________________________ Date ______________

Solve each problem below using at least two different approaches.

1. Jenny's great-grandmother is 90 years old. Jenny is 12 years old. What percent of Jenny's great-grandmother's age is Jenny's age?

2. Jenny's mom is 36 years old. What percent of Jenny's mother's age is Jenny's great-grandmother's age?

When solving real-world problems with percents, I can use the formula Quantity = Percent × Whole.

I am being asked to find the percent of the old number of participants. Therefore, the old number of participants is the whole.

1. The number of participants in the city choir decreased from 40 to 25.

   a. Express the new number of participants as a percent of the old number of participants.

   Let p represent the unknown percent.

$$\textbf{Quantity} = \textbf{Percent} \times \textbf{Whole}$$
$$25 = p \times 40$$
$$25\left(\frac{1}{40}\right) = p(40)\left(\frac{1}{40}\right)$$
$$0.625 = 1p$$
$$0.625 = p$$

The question is asking for a percent, so I need to convert this decimal to a percent.

In Lesson 2, I solved equations similar to this one.

   The new number of participants is 62.5% of the old number of participants.

The question has changed, and now the new number of participants is the whole.

   b. Express the old number of participants as a percent of the new number of participants.

   Let p represent the unknown percent.

$$\textbf{Quantity} = \textbf{Percent} \times \textbf{Whole}$$
$$40 = p \times 25$$
$$40\left(\frac{1}{25}\right) = p(25)\left(\frac{1}{25}\right)$$
$$1.6 = 1p$$
$$1.6 = p$$

   The old number of participants is 160% of the new number of participants.

The number of students who attend Berry is the whole, and the number of students who attend Newton is the quantity.

2. The number of students who attend Newton Elementary School is 75% of the number of students who attend Berry Middle School.

   a. Find the number of students who attend Newton if 500 students attend Berry.

   $\textbf{Quantity} = \textbf{Percent} \times \textbf{Whole}$ ***I know the percent and the whole; I have to determine the quantity.***

   Let n represent the number of students who attend Newton.

   $n = 75\%(500)$ ***Before solving the equation, I need to convert the percent to a decimal.***

   $n = 0.75(500)$ ***In order to solve for n, multiply the two factors together.***

   $n = 375$

   This means that 375 students attend Newton Elementary School.

   b. Find the number of students who attend Berry if 150 students attend Newton.

   $\textbf{Quantity} = \textbf{Percent} \times \textbf{Whole}$

   The question has changed. I now know the percent and the quantity. I need to calculate the whole.

   Let b represent the number of students who attend Berry.

$$
\begin{aligned}
150 &= 75\% \times b \\
150 &= 0.75b \\
150\left(\frac{1}{0.75}\right) &= 0.75b\left(\frac{1}{0.75}\right) \\
200 &= b
\end{aligned}
$$

   Therefore, 200 students attend Berry Middle School.

1. Solve each problem using an equation.
   a. 49.5 is what percent of 33?
   b. 72 is what percent of 180?
   c. What percent of 80 is 90?

2. This year, Benny is 12 years old, and his mom is 48 years old.
   a. What percent of his mom's age is Benny's age?
   b. What percent of Benny's age is his mom's age?
   c. In two years, what percent of his age will Benny's mom's age be at that time?
   d. In 10 years, what percent will Benny's mom's age be of his age?
   e. In how many years will Benny be 50% of his mom's age?
   f. As Benny and his mom get older, Benny thinks that the percent of difference between their ages will decrease as well. Do you agree or disagree? Explain your reasoning.

3. This year, Benny is 12 years old. His brother Lenny's age is 175% of Benny's age. How old is Lenny?

4. When Benny's sister Penny is 24, Benny's age will be 125% of her age.
   a. How old will Benny be then?
   b. If Benny is 12 years old now, how old is Penny now? Explain your reasoning.

5. Benny's age is currently 200% of his sister Jenny's age. What percent of Benny's age will Jenny's age be in 4 years?

6. At an animal shelter, there are 15 dogs, 12 cats, 3 snakes, and 5 parakeets.
   a. What percent of the number of cats is the number of dogs?
   b. What percent of the number of cats is the number of snakes?
   c. What percent less parakeets are there than dogs?
   d. Which animal has 80% of the number of another animal?
   e. Which animal makes up approximately 14% of the animals in the shelter?

7. Is 2 hours and 30 minutes more or less than 10% of a day? Explain your answer.

8. A club's membership increased from 25 to 30 members.
   a. Express the new membership as a percent of the old membership.
   b. Express the old membership as a percent of the new membership.

9. The number of boys in a school is 120% the number of girls at the school.
   a. Find the number of boys if there are 320 girls.
   b. Find the number of girls if there are 360 boys.

10. The price of a bicycle was increased from $\$300$ to $\$450$.
    a. What percent of the original price is the increased price?
    b. What percent of the increased price is the original price?

11. The population of Appleton is 175% of the population of Cherryton.
    a. Find the population in Appleton if the population in Cherryton is $4{,}000$ people.
    b. Find the population in Cherryton if the population in Appleton is $10{,}500$ people.

12. A statistics class collected data regarding the number of boys and the number of girls in each classroom at their school during homeroom. Some of their results are shown in the table below.

a. Complete the blank cells of the table using your knowledge about percent.

Number of Boys (x)	Number of Girls (y)	Number of Girls as a Percent of the Number of Boys
10	5	
	1	25%
18	12	
5	10	
4		50%
20		90%
	10	250%
	6	60%
11		200%
	5	$33\frac{1}{3}\%$
15		20%
	15	75%
6	18	
25	10	
10		110%
	2	10%
16		75%
	7	50%
3		200%
12	10	

b. Using a coordinate plane and grid paper, locate and label the points representing the ordered pairs (x, y).

c. Locate all points on the graph that would represent classrooms in which the number of girls y is 100% of the number of boys x. Describe the pattern that these points make.

d. Which points represent the classrooms in which the number of girls as a percent of the number of boys is greater than 100%? Which points represent the classrooms in which the number of girls as a percent of the number of boys is less than 100%? Describe the locations of the points in relation to the points in part (c).

e. Find three ordered pairs from your table representing classrooms where the number of girls is the same percent of the number of boys. Do these points represent a proportional relationship? Explain your reasoning.

f. Show the relationship(s) from part (e) on the graph, and label them with the corresponding equation(s).

g. What is the constant of proportionality in your equation(s), and what does it tell us about the number of girls and the number of boys at each point on the graph that represents it? What does the constant of proportionality represent in the table in part (a)?

Opening Exercise

Cassandra likes jewelry. She has five rings in her jewelry box.

a. In the box below, sketch Cassandra's five rings.

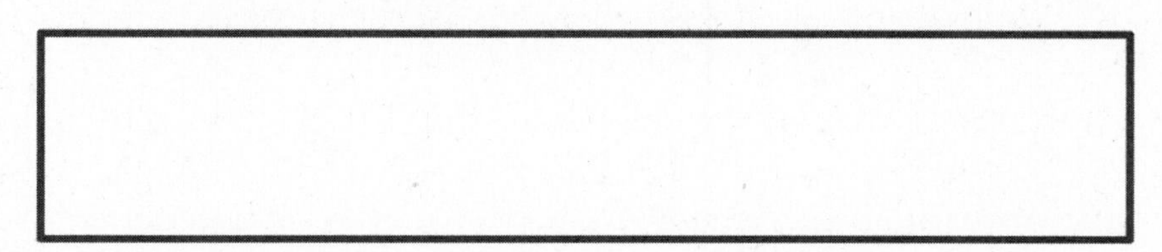

b. Draw a double number line diagram relating the number of rings as a percent of the whole set of rings.

c. What percent is represented by the whole collection of rings? What percent of the collection does each ring represent?

Example 1: Finding a Percent Increase

Cassandra's aunt said she will buy Cassandra another ring for her birthday. If Cassandra gets the ring for her birthday, what will be the percent increase in her ring collection?

Exercise 1

a. Jon increased his trading card collection by 5 cards. He originally had 15 cards. What is the percent increase? Use the equation $\text{Quantity} = \text{Percent} \times \text{Whole}$ to arrive at your answer, and then justify your answer using a numeric or visual model.

b. Suppose instead of increasing the collection by 5 cards, Jon increased his 15-card collection by just 1 card. Will the percent increase be the same as when Cassandra's ring collection increased by 1 ring (in Example 1)? Why or why not? Explain.

c. Based on your answer to part (b), how is displaying change as a percent useful?

Discussion

A sales representative is taking 10% off of your bill as an apology for any inconveniences.

Example 2: Percent Decrease

Ken said that he is going to reduce the number of calories that he eats during the day. Ken's trainer asked him to start off small and reduce the number of calories by no more than 7%. Ken estimated and consumed $2,200$ calories per day instead of his normal $2,500$ calories per day until his next visit with the trainer. Did Ken reduce his calorie intake by no more than 7%? Justify your answer.

Exercise 2

Skylar is answering the following math problem:

The value of an investment decreased by 10%. *The original amount of the investment was* $\$75.00$. *What is the current value of the investment?*

a. Skylar said 10% of $\$75.00$ is $\$7.50$, and since the investment decreased by that amount, you have to subtract $\$7.50$ from $\$75.00$ to arrive at the final answer of $\$67.50$. Create one algebraic equation that can be used to arrive at the final answer of $\$67.50$. Solve the equation to prove it results in an answer of $\$67.50$. Be prepared to explain your thought process to the class.

b. Skylar wanted to show the proportional relationship between the dollar value of the original investment, x, and its value after a 10% decrease, y. He creates the table of values shown. Does it model the relationship? Explain. Then, provide a correct equation for the relationship Skylar wants to model.

x	y
75	7.5
100	10
200	20
300	30
400	40

Example 3: Finding a Percent Increase or Decrease

Justin earned 8 badges in Scouts as of the Scout Master's last report. Justin wants to complete 2 more badges so that he will have a total of 10 badges earned before the Scout Master's next report.

a. If Justin completes the additional 2 badges, what will be the percent increase in badges?

b. Express the 10 badges as a percent of the 8 badges.

c. Does 100% plus your answer in part (a) equal your answer in part (b)? Why or why not?

Example 4: Finding the Original Amount Given a Percent Increase or Decrease

The population of cats in a rural neighborhood has declined in the past year by roughly 30%. Residents hypothesize that this is due to wild coyotes preying on the cats. The current cat population in the neighborhood is estimated to be 12. Approximately how many cats were there originally?

Example 5: Finding the Original Amount Given a Percent Increase or Decrease

Lu's math score on her achievement test in seventh grade was a 650. Her math teacher told her that her test level went up by 25% from her sixth grade test score level. What was Lu's test score level in sixth grade?

Closing

Phrase	Whole Unit (100%)
"Mary has 20% more money than John."	
"Anne has 15% less money than John."	
"What percent more (money) does Anne have than Bill?"	
"What percent less (money) does Bill have than Anne?"	

Lesson Summary

- Within each problem, there are keywords that determine if the problem represents a percent increase or a percent decrease.
- Equations can be used to solve percent problems using the basic equation
$$\text{Quantity} = \text{Percent} \times \text{Whole}.$$
- *Quantity* in the percent formula is the amount of change (increase or decrease) or the amount after the change.
- *Whole* in the percent formula represents the original amount.

Name ______________________________ Date ______________

Erin wants to raise her math grade to a 95 to improve her chances of winning a math scholarship. Her math average for the last marking period was an 81. Erin decides she must raise her math average by 15% to meet her goal. Do you agree? Why or why not? Support your written answer by showing your math work.

> This is the original price, so it represents the whole.

1. A store is advertising 20% off a new Blu-ray Player that regularly sells for $60.

   a. What is the sale price of the item?

   $100\% - 20\% = 80\%$

   Therefore, I am paying $\mathbf{80\%}$ ***of the original price.***

   > In order to solve this problem, I determine the percent of the original price I have to pay.

   Let n represent the sale price.

   $n = 80\% \times 60$

   $n = 0.8(60)$

   $n = 48$

   The sale price is $\mathbf{\$48}$.

   > Percent increase and decrease problems are still percent problems in the real-world, so I use the formula
   > $\text{Quantity} = \text{Percent} \times \text{Whole}$.

   b. If 6% sales tax is charged on the sale price, what is the total with tax?

   $100\% + 6\% = 106\%$

   I pay $\mathbf{106\%}$ ***of the sale price ($48), which will represent the whole since I am finding the sale price with tax.***

   > Sales tax increases the price, so I have to pay 100% of the sale price, plus the extra 6% for tax.

   Let t represent the sale price with tax.

   $t = 106\% \times 48$

   $t = 1.06 \times 48$

   $t = 50.88$

   The sale price with tax is $\mathbf{\$50.88}$.

2. The Parent-Teacher Organization had 30 participants attend the first meeting of the school year and only 24 participants attend the second meeting. Find the percent decrease in the participants from the first meeting to the second meeting.

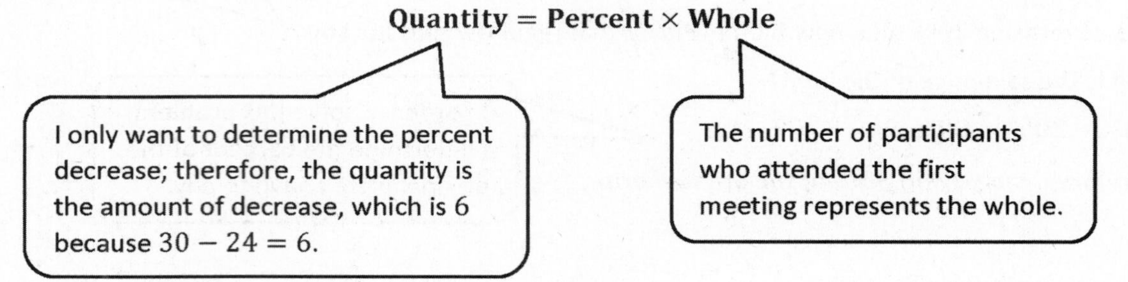

Let p represent the percent decrease.

$$6 = p \times 30$$
$$6\left(\frac{1}{30}\right) = p \times 30\left(\frac{1}{30}\right)$$
$$0.2 = p$$
$$20\% = p$$

Therefore, the percent decrease is 20%.

3. Chelsey is keeping a diary to keep track of the number of days she goes running. In the first 40 days, she ran 40% of the days. She kept recording for another 20 days and then found that the total number of days she ran increased to 50%. How many of the final 20 days did Chelsey go running?

Let r represent the number of days Chelsey ran during the first 40 days.

$r = 40\%(40)$
$r = (0.40)(40)$
$r = 16$

I know I can find the number of days Chelsey ran during the first 40 days where 40 represents the whole.

Chelsey ran 16 days in the first 40 days.

Let d represent the total number of days Chelsey went running.

$d = 50\%(60)$
$d = 0.50(60)$
$d = 30$

$40 + 20 = 60$ This means that Chelsey kept the diary for a total of 60 days.

Chelsey ran a total of 30 days.

$30 - 16 = 14$

I know the number of days Chelsey ran during the last 20 days is the difference between the total number of days she ran and the number of days she ran during the first 40 days.

Chelsey ran 14 days during the last 20 days.

1. A store advertises 15% off an item that regularly sells for $300.
   a. What is the sale price of the item?
   b. How is a 15% discount similar to a 15% decrease? Explain.
   c. If 8% sales tax is charged on the sale price, what is the total with tax?
   d. How is 8% sales tax like an 8% increase? Explain.

2. An item that was selling for $72.00 is reduced to $60.00. Find the percent decrease in price. Round your answer to the nearest tenth.

3. A baseball team had 80 players show up for tryouts last year and this year had 96 players show up for tryouts. Find the percent increase in players from last year to this year.

4. At a student council meeting, there was a total of 60 students present. Of those students, 35 were female.
   a. By what percent is the number of females greater than the number of males?
   b. By what percent is the number of males less than the number of females?
   c. Why is the percent increase and percent decrease in parts (a) and (b) different?

5. Once each day, Darlene writes in her personal diary and records whether the sun is shining or not. When she looked back though her diary, she found that over a period of 600 days, the sun was shining 60% of the time. She kept recording for another 200 days and then found that the total number of sunny days dropped to 50%. How many of the final 200 days were sunny days?

6. Henry is considering purchasing a mountain bike. He likes two bikes: One costs $500, and the other costs $600. He tells his dad that the bike that is more expensive is 20% more than the cost of the other bike. Is he correct? Justify your answer.

7. State two numbers such that the lesser number is 25% less than the greater number.

8. State two numbers such that the greater number is 75% more than the lesser number.

9. Explain the difference in your thought process for Problems 7 and 8. Can you use the same numbers for each problem? Why or why not?

10. In each of the following expressions, c represents the original cost of an item.

    i. $0.90c$

    ii. $0.10c$

    iii. $c - 0.10c$

    a. Circle the expression(s) that represents 10% of the original cost. If more than one answer is correct, explain why the expressions you chose are equivalent.

    b. Put a box around the expression(s) that represents the final cost of the item after a 10% decrease. If more than one is correct, explain why the expressions you chose are equivalent.

    c. Create a word problem involving a percent decrease so that the answer can be represented by expression (ii).

    d. Create a word problem involving a percent decrease so that the answer can be represented by expression (i).

    e. Tyler wants to know if it matters if he represents a situation involving a 25% decrease as $0.25x$ or $(1 - 0.25)x$. In the space below, write an explanation that would help Tyler understand how the context of a word problem often determines how to represent the situation.

Opening Exercise

What are the whole number factors of 100? What are the multiples of those factors? How many multiples are there of each factor (up to 100)?

Factors of 100	Multiples of the Factors of 100	Number of Multiples
100	100	1
50	$50, 100$	2
1	$1, 2, 3, 4, 5, 6, \ldots, 98, 99, 100$	100

Example 1: Using a Modified Double Number Line with Percents

The 42 students who play wind instruments represent 75% of the students who are in band. How many students are in band?

Exercises 1–3

1. Bob's Tire Outlet sold a record number of tires last month. One salesman sold 165 tires, which was 60% of the tires sold in the month. What was the record number of tires sold?

2. Nick currently has $7{,}200$ points in his fantasy baseball league, which is 20% more points than Adam. How many points does Adam have?

3. Kurt has driven 276 miles of his road trip but has 70% of the trip left to go. How many more miles does Kurt have to drive to get to his destination?

Example 2: Mental Math Using Factors of 100

Answer each part below using only mental math, and describe your method.

a. If 39 is 1% of a number, what is that number? How did you find your answer?

b. If 39 is 10% of a number, what is that number? How did you find your answer?

c. If 39 is 5% of a number, what is that number? How did you find your answer?

d. If 39 is 15% of a number, what is that number? How did you find your answer?

e. If 39 is 25% of a number, what is that number? How did you find your answer?

Exercises 4–5

4. Derrick had a 0.250 batting average at the end of his last baseball season, which means that he got a hit 25% of the times he was up to bat. If Derrick had 47 hits last season, how many times did he bat?

5. Nelson used 35% of his savings account for his class trip in May. If he used $\$140$ from his savings account while on his class trip, how much money was in his savings account before the trip?

Lesson Summary

To find 100% of the whole, you can use a variety of methods, including factors of 100 (1, 2, 4, 5, 10, 20, 25, 50, and 100) and double number lines. Both methods will require breaking 100% into equal-sized intervals. Use the greatest common factor of 100 and the percent corresponding to the part.

Name ______________________________ Date ________________

1. A tank that is 40% full contains 648 gallons of water. Use a double number line to find the maximum capacity of the water tank.

2. Loretta picks apples for her grandfather to make apple cider. She brings him her cart with 420 apples. Her grandfather smiles at her and says, "Thank you, Loretta. That is 35% of the apples that we need."

   Use mental math to find how many apples Loretta's grandfather needs. Describe your method.

1. The number of students who attended the spring school dance was a 20% decrease from the number of students who attended the fall dance. If 424 students attended the spring dance, how many students attended the fall dance?

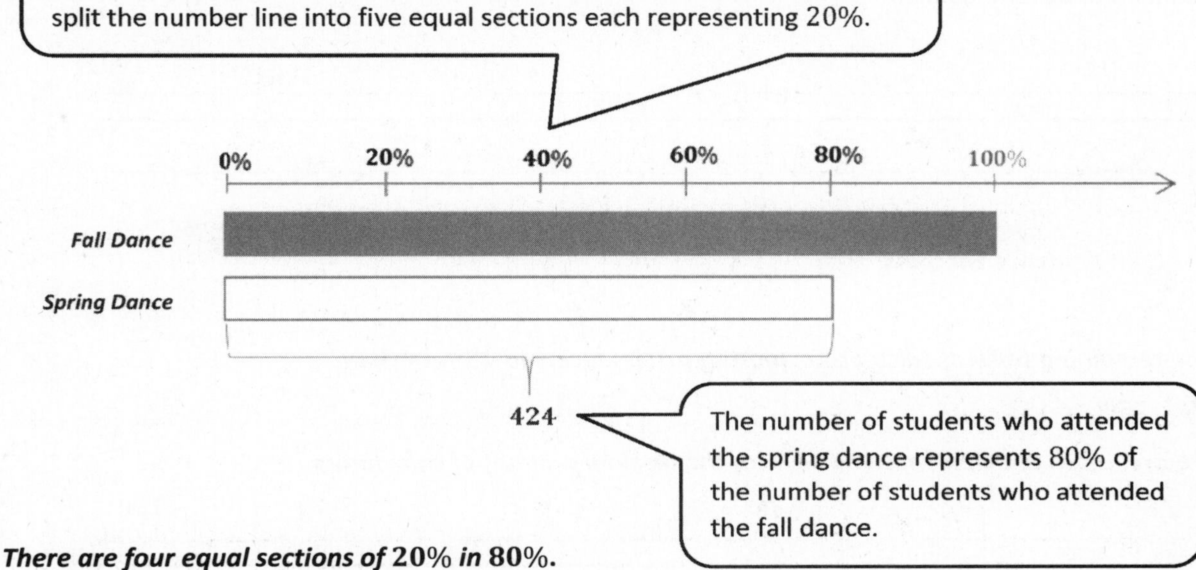

There are four equal sections of $\mathbf{20\%}$ ***in*** $\mathbf{80\%}$.

$\mathbf{424 \div 4 = 106}$

Therefore, each of these sections would represent $\mathbf{106}$.

$\mathbf{106 \times 5 = 530}$

There are five equal sections of $\mathbf{20\%}$ ***in*** $\mathbf{100\%}$***. I know that each section of*** $\mathbf{20\%}$ ***represents*** $\mathbf{106}$ ***students. The number of students who attended the fall dance was*** $\mathbf{530}$.

Use any method to solve the problem.

2. A middle school ordered new calculators. The science teachers received 40% of the calculators, and the math teachers received 75% of the remaining calculators. There were 60 calculators that were not given to either science or math teachers. How many calculators were given to the science teachers? Math teachers? How many calculators were originally ordered?

I know that 40% of the calculators were given to science teachers, which means 60% of the calculators were given to other teachers.

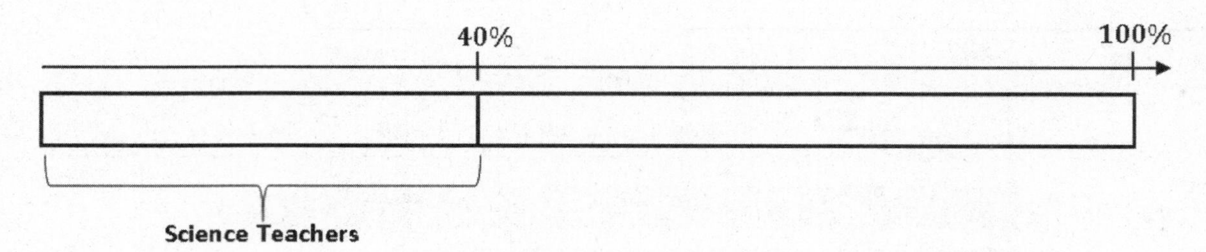

Of the remaining 60% of calculators, math teachers received 75% of them.

$60\% \times 75\% = 45\%$

Therefore, math teachers received 45% of the original amount of calculators.

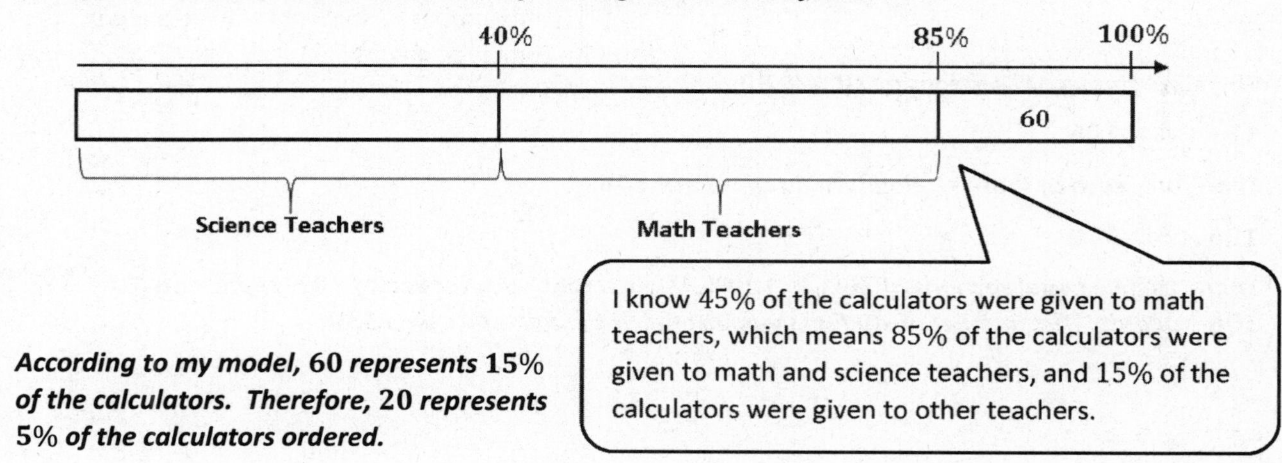

According to my model, 60 represents 15% of the calculators. Therefore, 20 represents 5% of the calculators ordered.

If I know 20 represents 5%, I can multiply both values by eight in order to determine how many calculators represents 40% of the calculators ordered. For the same reason, I need to multiply both values by nine to determine the number of calculators the math teachers received.

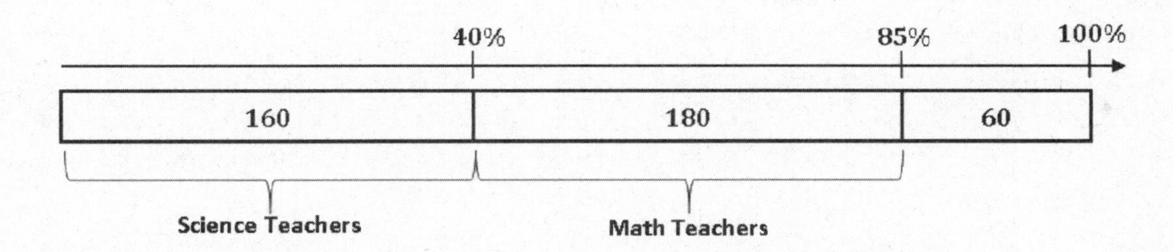

Science teachers received* 160 *calculators. Math teachers received* 180 *calculators.

$160 + 180 + 60 = 400$

The school originally ordered* 400 *calculators.

Use a double number line to answer Problems 1–5.

1. Tanner collected 360 cans and bottles while fundraising for his baseball team. This was 40% of what Reggie collected. How many cans and bottles did Reggie collect?

2. Emilio paid $287.50 in taxes to the school district that he lives in this year. This year's taxes were a 15% increase from last year. What did Emilio pay in school taxes last year?

3. A snowmobile manufacturer claims that its newest model is 15% lighter than last year's model. If this year's model weighs 799 lb., how much did last year's model weigh?

4. Student enrollment at a local school is concerning the community because the number of students has dropped to 504, which is a 20% decrease from the previous year. What was the student enrollment the previous year?

5. The color of paint used to paint a race car includes a mixture of yellow and green paint. Scotty wants to lighten the color by increasing the amount of yellow paint 30%. If a new mixture contains 3.9 liters of yellow paint, how many liters of yellow paint did he use in the previous mixture?

Use factors of 100 and mental math to answer Problems 6–10. Describe the method you used.

6. Alexis and Tasha challenged each other to a typing test. Alexis typed 54 words in one minute, which was 120% of what Tasha typed. How many words did Tasha type in one minute?

7. Yoshi is 5% taller today than she was one year ago. Her current height is 168 cm. How tall was she one year ago?

8. Toya can run one lap of the track in 1 min. 3 sec., which is 90% of her younger sister Niki's time. What is Niki's time for one lap of the track?

9. An animal shelter houses only cats and dogs, and there are 25% more cats than dogs. If there are 40 cats, how many dogs are there, and how many animals are there total?

10. Angie scored 91 points on a test but only received a 65% grade on the test. How many points were possible on the test?

For Problems 11–17, find the answer using any appropriate method.

11. Robbie owns 15% more movies than Rebecca, and Rebecca owns 10% more movies than Joshua. If Rebecca owns 220 movies, how many movies do Robbie and Joshua each have?

12. 20% of the seventh-grade students have math class in the morning. $16\frac{2}{3}$% of those students also have science class in the morning. If 30 seventh-grade students have math class in the morning but not science class, find how many seventh-grade students there are.

13. The school bookstore ordered three-ring notebooks. They put 75% of the order in the warehouse and sold 80% of the rest in the first week of school. There are 25 notebooks left in the store to sell. How many three-ring notebooks did they originally order?

14. In the first game of the year, the modified basketball team made 62.5% of their foul shot free throws. Matthew made all 6 of his free throws, which made up 25% of the team's free throws. How many free throws did the team miss altogether?

15. Aiden's mom calculated that in the previous month, their family had used 40% of their monthly income for gasoline, and 63% of that gasoline was consumed by the family's SUV. If the family's SUV used $261.45 worth of gasoline last month, how much money was left after gasoline expenses?

16. Rectangle A is a scale drawing of Rectangle B and has 25% of its area. If Rectangle A has side lengths of 4 cm and 5 cm, what are the side lengths of Rectangle B?

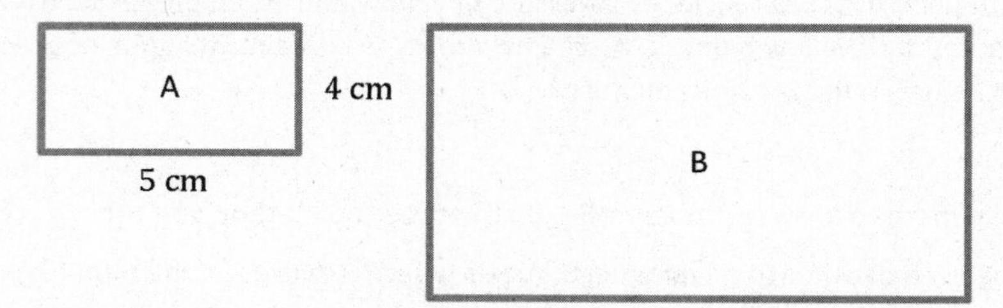

17. Ted is a supervisor and spends 20% of his typical work day in meetings and 20% of that meeting time in his daily team meeting. If he starts each day at 7:30 a.m., and his daily team meeting is from 8:00 a.m. to 8:20 a.m., when does Ted's typical work day end?

Number Correct: ______

Percent More or Less—Round 1

Directions: Find each missing value.

1.	100% of 10 is ___?	
2.	10% of 10 is ___?	
3.	10% more than 10 is ___?	
4.	11 is ___% more than 10?	
5.	11 is ___% of 10?	
6.	11 is 10% more than ___?	
7.	110% of 10 is ___?	
8.	10% less than 10 is ___?	
9.	9 is ___% less than 10?	
10.	9 is ___% of 10?	
11.	9 is 10% less than ___?	
12.	10% of 50 is ___?	
13.	10% more than 50 is ___?	
14.	55 is ___% of 50?	
15.	55 is ___% more than 50?	
16.	55 is 10% more than ___?	
17.	110% of 50 is ___?	
18.	10% less than 50 is ___?	
19.	45 is ___% of 50?	
20.	45 is ___% less than 50?	
21.	45 is 10% less than ___?	
22.	40 is ___% less than 50?	

23.	15% of 80 is ___?	
24.	15% more than 80 is ___?	
25.	What is 115% of 80?	
26.	92 is 115% of ___?	
27.	92 is ___% more than 80?	
28.	115% of 80 is ___?	
29.	What is 15% less than 80?	
30.	What % of 80 is 68?	
31.	What % less than 80 is 68?	
32.	What % less than 80 is 56?	
33.	What % of 80 is 56?	
34.	What is 20% more than 50?	
35.	What is 30% more than 50?	
36.	What is 140% of 50?	
37.	What % of 50 is 85?	
38.	What % more than 50 is 85?	
39.	What % less than 50 is 35?	
40.	What % of 50 is 35?	
41.	1 is what % of 50?	
42.	6 is what % of 50?	
43.	24% of 50 is?	
44.	24% more than 50 is ___?	

Number Correct: ______
Improvement: ______

Percent More or Less—Round 2

Directions: Find each missing value.

1.	100% of 20 is __?		23.	15% of 60 is __?	
2.	10% of 20 is __?		24.	15% more than 60 is __?	
3.	10% more than 20 is __?		25.	What is 115% of 60?	
4.	22 is __% more than 20?		26.	69 is 115% of __?	
5.	22 is __% of 20?		27.	69 is __% more than 60?	
6.	22 is 10% more than __?		28.	115% of 60 is __?	
7.	110% of 20 is __?		29.	What is 15% less than 60?	
8.	10% less than 20 is __?		30.	What % of 60 is 51?	
9.	18 is __% less than 20?		31.	What % less than 60 is 51?	
10.	18 is __% of 20?		32.	What % less than 60 is 42?	
11.	18 is 10% less than __?		33.	What % of 60 is 42?	
12.	10% of 200 is __?		34.	What is 20% more than 80?	
13.	10% more than 200 is __?		35.	What is 30% more than 80?	
14.	220 is __% of 200?		36.	What is 140% of 80?	
15.	220 is __% more than 200?		37.	What % of 80 is 104?	
16.	220 is 10% more than __?		38.	What % more than 80 is 104?	
17.	110% of 200 is __?		39.	What % less than 80 is 56?	
18.	10% less than 200 is __?		40.	What % of 80 is 56?	
19.	180 is __% of 200?		41.	1 is what % of 200?	
20.	180 is __% less than 200?		42.	6 is what % of 200?	
21.	180 is 10% less than __?		43.	24% of 200 is ___?	
22.	160 is __% less than 200?		44.	24% more than 200 is ___?	

Opening Exercise

Solve the following problem using mental math only. Be prepared to discuss your method with your classmates.

Cory and Everett have collected model cars since the third grade. Cory has 80 model cars in his collection, which is 25% more than Everett has. How many model cars does Everett have?

Example 1: Mental Math and Percents

a. 75% of the students in Jesse's class are 60 inches or taller. If there are 20 students in her class, how many students are 60 inches or taller?

b. Bobbie wants to leave a tip for her waitress equal to 15% of her bill. Bobbie's bill for her lunch is $18. How much money represents 15% of the bill?

Exercises

1. Express 9 hours as a percentage of 3 days.

2. Richard works from 11:00 a.m. to 3:00 a.m. His dinner break is 75% of the way through his work shift. What time is Richard's dinner break?

3. At a playoff basketball game, there were 370 fans cheering for school A and 555 fans cheering for school B.
   a. Express the number of fans cheering for school A as a percent of the number of fans cheering for school B.

b. Express the number of fans cheering for school B as a percent of the number of fans cheering for school A.

c. What percent more fans were there for school B than for school A?

4. Rectangle A has a width of 8 cm and a length of 16 cm. Rectangle B has the same area as the first, but its width is 62.5% of the width of the first rectangle. Express the length of Rectangle B as a percent of the length of Rectangle A. What percent more or less is the length of Rectangle B than the length of Rectangle A?

5. A plant in Mikayla's garden was 40 inches tall one day and was 4 feet tall one week later. By what percent did the plant's height increase over one week?

6. Loren must obtain a minimum number of signatures on a petition before it can be submitted. She was able to obtain 672 signatures, which is 40% more than she needs. How many signatures does she need?

Lesson Summary

- Identify the type of percent problem that is being asked as a comparison of quantities or a part of a whole.
- Identify what numbers represent the part, the whole, and the percent, and use the representation

$$\text{Quantity} = \text{Percent} \times \text{Whole}.$$

- A strategy to solving percents using mental math is to rewrite a percent using 1%, 5%, or 10%. These percents can be solved mentally. For example: $13\% = 10\% + 3(1\%)$. To find 13% of 70, find 10% of 70 as 7, 1% of 70 as 0.7, so 13% of 70 is $7 + 3(0.7) = 7 + 2.10 = 9.10$.

Name ______________________________ Date ______________

1. Parker was able to pay for 44% of his college tuition with his scholarship. The remaining $\$10{,}054.52$ he paid for with a student loan. What was the cost of Parker's tuition?

2. Two bags contain marbles. Bag A contains 112 marbles, and Bag B contains 140 marbles. What percent fewer marbles does Bag A have than Bag B?

3. There are 42 students on a large bus, and the rest are on a smaller bus. If 40% of the students are on the smaller bus, how many total students are on the two buses?

1. Monroe Middle School has 564 students, which is 60% of the number of students who attend Wilson Middle School. How many student attend Wilson Middle School?

$$\textbf{Quantity} = \textbf{Percent} \times \textbf{Whole}$$

Let s represent the number of students who attend Wilson Middle School.

$$564 = 60\% \times s$$
$$564 = 0.6 \times s$$
$$\left(\frac{1}{0.6}\right)(564) = \left(\frac{1}{0.6}\right)(0.6)s$$
$$940 = s$$

I know the number of students who attend Wilson Middle School represents the whole because it is the value being compared to the percent.

940 ***students attend Wilson Middle School.***

2. In a school-wide survey, students could either choose basketball or soccer as their favorite sport. The number of people who chose basketball as their favorite sport was 20% greater than the number of people who chose soccer. If 450 students chose basketball as their favorite sport, how many people took the survey?

Let s represent the number of people who chose soccer as their favorite sport.

$$450 = 120\% \times s$$
$$450 = 1.2s$$
$$\left(\frac{1}{1.2}\right)(450) = \left(\frac{1}{1.2}\right)(1.2s)$$
$$375 = s$$

The number of people who chose basketball is 120% of those who chose soccer because the number of people who chose basketball is equal to the number of people who chose soccer plus an extra 20% of the people.

The number of people who chose soccer as their favorite sport is 375.

$375 + 450 = 825$

The total number of people who completed the survey was 825.

3. Electricity City increases the price of televisions by 25%. If a television now sells for $425, what was the original price?

$$\text{Quantity} = \text{Percent} \times \text{Whole}$$

Let p represent the original price, in dollars, of the television.

$$425 = 125\% \times p$$
$$425 = 1.25p$$
$$\left(\frac{1}{1.25}\right)(425) = \left(\frac{1}{1.25}\right)(1.25)p$$
$$340 = p$$

I know the television sold for 125% of the original price because the cost would be 100% of the original price plus the 25% increase.

The original price of the television is $\$340$.

I can do a quick check to make sure my answer makes sense. I know that my answer is the original price, which means it should be a smaller value than the price the television sold for.

4. Christopher spent 20% of his paycheck at the mall, and 35% of that amount was spent on a new video game. Christopher spent a total of $21.35 on the video game. How much money was in Christopher's paycheck?

Let m represent the amount Christopher spent at the mall.

$$21.35 = \frac{35}{100}m$$
$$\left(\frac{100}{35}\right)(21.35) = \left(\frac{100}{35}\right)\left(\frac{35}{100}m\right)$$
$$61 = m$$

The amount Christopher spent at the mall is the whole, and I can determine the amount he spent on a video game.

Christopher spent $\$61.00$ at the mall. Let p represent the amount of money in Christopher's paycheck.

I can use the information I found in the previous equation to determine the amount of money in Christopher's paycheck.

$$61 = \frac{20}{100}p$$
$$\left(\frac{100}{20}\right)(61) = \left(\frac{100}{20}\right)\left(\frac{20}{100}p\right)$$
$$305 = p$$

Christopher's paycheck was $\$305.00$.

1. Micah has 294 songs stored in his phone, which is 70% of the songs that Jorge has stored in his phone. How many songs are stored on Jorge's phone?

2. Lisa sold 81 magazine subscriptions, which is 27% of her class's fundraising goal. How many magazine subscriptions does her class hope to sell?

3. Theresa and Isaiah are comparing the number of pages that they read for pleasure over the summer. Theresa read $2{,}210$ pages, which was 85% of the number of pages that Isaiah read. How many pages did Isaiah read?

4. In a parking garage, the number of SUVs is 40% greater than the number of non-SUVs. Gina counted 98 SUVs in the parking garage. How many vehicles were parked in the garage?

5. The price of a tent was decreased by 15% and sold for $\$76.49$. What was the original price of the tent in dollars?

6. 40% of the students at Rockledge Middle School are musicians. 75% of those musicians have to read sheet music when they play their instruments. If 38 of the students can play their instruments without reading sheet music, how many students are there at Rockledge Middle School?

7. At Longbridge Middle School, 240 students said that they are an only child, which is 48% of the school's student enrollment. How many students attend Longbridge Middle School?

8. Grace and her father spent $4\frac{1}{2}$ hours over the weekend restoring their fishing boat. This time makes up 6% of the time needed to fully restore the boat. How much total time is needed to fully restore the boat?

9. Bethany's mother was upset with her because Bethany's text messages from the previous month were 218% of the amount allowed at no extra cost under her phone plan. Her mother had to pay for each text message over the allowance. Bethany had $5{,}450$ text messages last month. How many text messages is she allowed under her phone plan at no extra cost?

10. Harry used 84% of the money in his savings account to buy a used dirt bike that cost him $\$1{,}050$. How much money is left in Harry's savings account?

11. 15% of the students in Mr. Riley's social studies classes watch the local news every night. Mr. Riley found that 136 of his students do not watch the local news. How many students are in Mr. Riley's social studies classes?

12. Grandma Bailey and her children represent about 9.1% of the Bailey family. If Grandma Bailey has 12 children, how many members are there in the Bailey family?

13. Shelley earned 20% more money in tips waitressing this week than last week. This week she earned $\$72.00$ in tips waitressing. How much money did Shelley earn last week in tips?

14. Lucy's savings account has 35% more money than her sister Edy's. Together, the girls have saved a total of \$206.80. How much money has each girl saved?

15. Bella spent 15% of her paycheck at the mall, and 40% of that was spent at the movie theater. Bella spent a total of \$13.74 at the movie theater for her movie ticket, popcorn, and a soft drink. How much money was in Bella's paycheck?

16. On a road trip, Sara's brother drove 47.5% of the trip, and Sara drove 80% of the remainder. If Sara drove for 4 hours and 12 minutes, how long was the road trip?

Example 1: A Video Game Markup

Games Galore Super Store buys the latest video game at a wholesale price of $\$30.00$. The markup rate at Game's Galore Super Store is 40%. You use your allowance to purchase the game at the store. How much will you pay, not including tax?

a. Write an equation to find the price of the game at Games Galore Super Store. Explain your equation.

b. Solve the equation from part (a).

c. What was the total markup of the video game? Explain.

d. You and a friend are discussing markup rate. He says that an easier way to find the total markup is by multiplying the wholesale price of $\$30.00$ by 40%. Do you agree with him? Why or why not?

Example 2: Black Friday

A $300 mountain bike is discounted by 30% and then discounted an additional 10% for shoppers who arrive before 5:00 a.m.

a. Find the sales price of the bicycle.

b. In all, by how much has the bicycle been discounted in dollars? Explain.

c. After both discounts were taken, what was the total percent discount?

d. Instead of purchasing the bike for $300, how much would you save if you bought it before 5:00 a.m.?

Exercises 1–3

1. Sasha went shopping and decided to purchase a set of bracelets for 25% off the regular price. If Sasha buys the bracelets today, she will save an additional 5%. Find the sales price of the set of bracelets with both discounts. How much money will Sasha save if she buys the bracelets today?

2. A golf store purchases a set of clubs at a wholesale price of $250. Mr. Edmond learned that the clubs were marked up 200%. Is it possible to have a percent increase greater than 100%? What is the retail price of the clubs?

3. Is a percent increase of a set of golf clubs from $250 to $750 the same as a markup rate of 200%? Explain.

Example 3: Working Backward

A car that normally sells for $20,000 is on sale for $16,000. The sales tax is 7.5%.

a. What percent of the original price of the car is the final price?

b. Find the discount rate.

c. By law, sales tax has to be applied to the discount price. However, would it be better for the consumer if the 7.5% sales tax was calculated before the 20% discount was applied? Why or why not?

d. Write an equation applying the commutative property to support your answer to part (c).

Exercise 4

a. Write an equation to determine the selling price in dollars, p, on an item that is originally priced s dollars after a markup of 25%.

b. Create and label a table showing five possible pairs of solutions to the equation.

c. Create and label a graph of the equation.

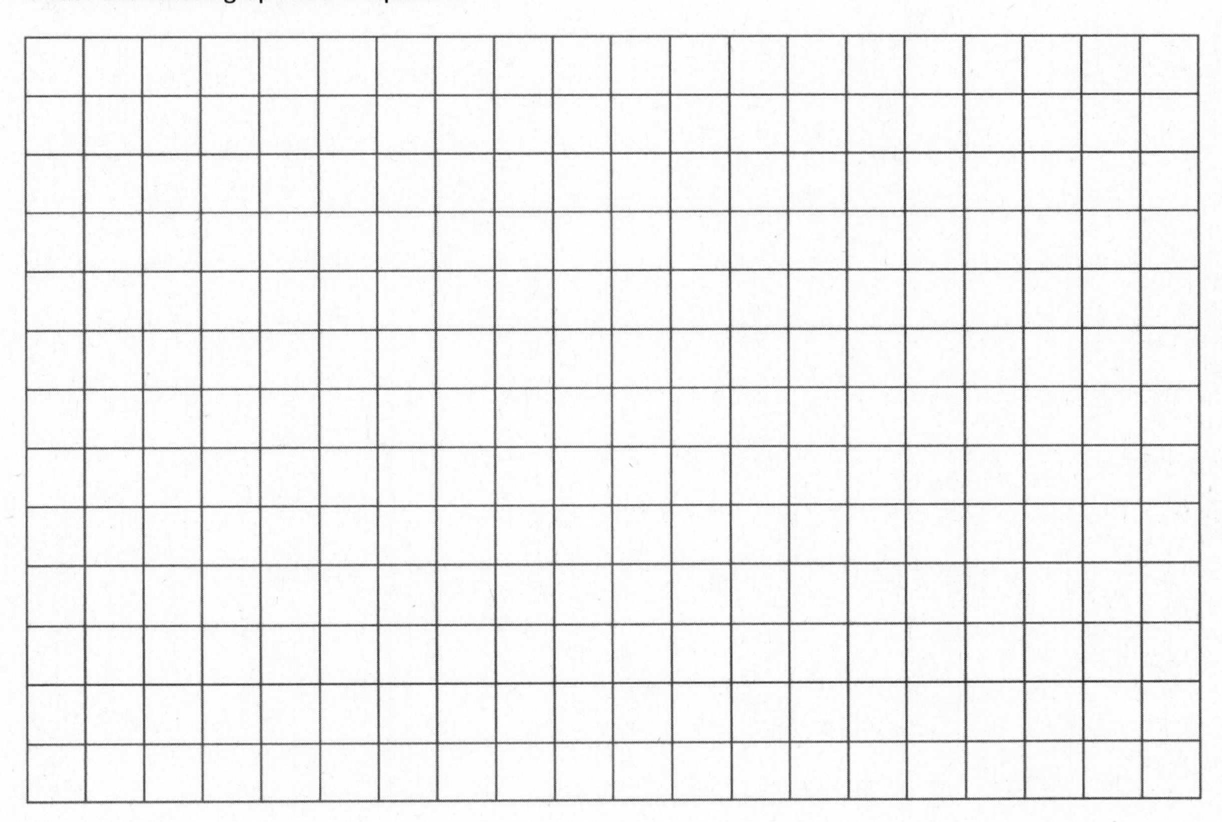

d. Interpret the points $(0,0)$ and $(1,r)$.

Exercise 5

Use the following table to calculate the markup or markdown rate. Show your work. Is the relationship between the original price and the selling price proportional or not? Explain.

Original Price, m (in dollars)	**Selling Price, p (in dollars)**
$1{,}750$	$1{,}400$
$1{,}500$	$1{,}200$
$1{,}250$	$1{,}000$
$1{,}000$	800
750	600

Lesson Summary

- To find the final price after a markup or markdown, multiply the whole by $(1 \pm m)$, where m is the markup/markdown rate.
- To apply multiple discount rates to the price of an item, you must find the first discount price and then use this answer to get the second discount price.

Name ______________________________ Date ______________

A store that sells skis buys them from a manufacturer at a wholesale price of $57. The store's markup rate is 50%.

a. What price does the store charge its customers for the skis?

b. What percent of the original price is the final price? Show your work.

c. What is the percent increase from the original price to the final price?

1. A school is conducting a fundraiser by selling sweatshirts. The school marks up the price of the sweatshirts by 40% in order to make a large profit. The school sells each sweatshirt for $\$28$.

   a. What is the original cost of the sweatshirts?

   $\textbf{Selling Price} = \mathbf{140\%(Original\ Cost)}$

   > There was a markup of 40%; therefore, the selling price is 140% of the original price because I have to pay 100% of the original price and the 40% markup.

   Let c represent the original cost.

   $$\mathbf{28 = 140\%(c)}$$
   $$\mathbf{28 = 1.4c}$$
   $$\mathbf{\frac{1}{1.4}(28) = \left(\frac{1}{1.4}\right)(1.4)(c)}$$
   $$\mathbf{20 = c}$$

   Therefore, the original cost of each sweatshirt was $\mathbf{\$20}$**.**

   b. How much did the school earn on each sweatshirt due to the markup?

   $\mathbf{\$28 - \$20 = \$8}$

   The school earned $\mathbf{\$8}$ ***due to the markup.***

> Due to the discount, I will pay $100\% - 25\%$, or 75%, of the original price. However, the sales tax will force me to pay $100\% + 6\%$, or 106%, of the discounted price.

2. A tool bench costs $\$650$ but is marked 25% off. If sales tax is 6%, what is the final cost of the tool bench?

   Let c represent the final cost of the tool bench.

   $$\mathbf{c = (original\ cost)(percentage\ paid)(sales\ tax)}$$
   $$\mathbf{c = (650)(0.75)(1.06)}$$
   $$\mathbf{c = 516.75}$$

   > The commutative and associative properties can help me calculate the product.

   Therefore, the final price of the tool bench is $\mathbf{\$516.75}$**.**

3. A local store sells a small television for $185. However, the local store buys the same television from a wholesaler for $100. What is the markup rate?

Let p represent the percent of the original price.

> The selling price is $185, and the original price is $100.

$$\text{Selling Price} = \text{Percent} \times \text{Original Price}$$
$$185 = p \times 100$$
$$\left(\frac{1}{100}\right)(185) = \left(\frac{1}{100}\right)(p \times 100)$$
$$1.85 = p$$

> This means that the selling price is 185% of the wholesale price, but this is not the markup rate.

$185\% - 100\% = 85\%$

The markup rate is 85%.

4. The sale price for a computer is $450. The original price was first discounted by 25% and then discounted an additional 20%. Find the original price of the computer.

Let c represent the cost of the computer before the additional 20% discount.

> I need to work backward and first find the cost of the computer before the second discount.

$$450 = 0.80c$$
$$\left(\frac{1}{0.80}\right)(450) = \left(\frac{1}{0.80}\right)(0.80c)$$
$$562.5 = c$$

The cost of the computer before the second discount was $562.50.

> Now, I can use the price of the computer after the first discount to determine the original cost.

Let p represent the original cost of the computer.

> A discount of 25% means I pay 75% of the original price.

$$562.5 = 0.75p$$
$$\left(\frac{1}{0.75}\right)(562.5) = \left(\frac{1}{0.75}\right)(0.75p)$$
$$750 = p$$

The original cost of the computer was $750.

1. You have a coupon for an additional 25% off the price of any sale item at a store. The store has put a robotics kit on sale for 15% off the original price of $\$40$. What is the price of the robotics kit after both discounts?

2. A sign says that the price marked on all music equipment is 30% off the original price. You buy an electric guitar for the sale price of $\$315$.
   a. What is the original price?
   b. How much money did you save off the original price of the guitar?
   c. What percent of the original price is the sale price?

3. The cost of a New York Yankee baseball cap is $\$24.00$. The local sporting goods store sells it for $\$30.00$. Find the markup rate.

4. Write an equation to determine the selling price in dollars, p, on an item that is originally priced s dollars after a markdown of 15%.
   a. Create and label a table showing five possible pairs of solutions to the equation.
   b. Create and label a graph of the equation.

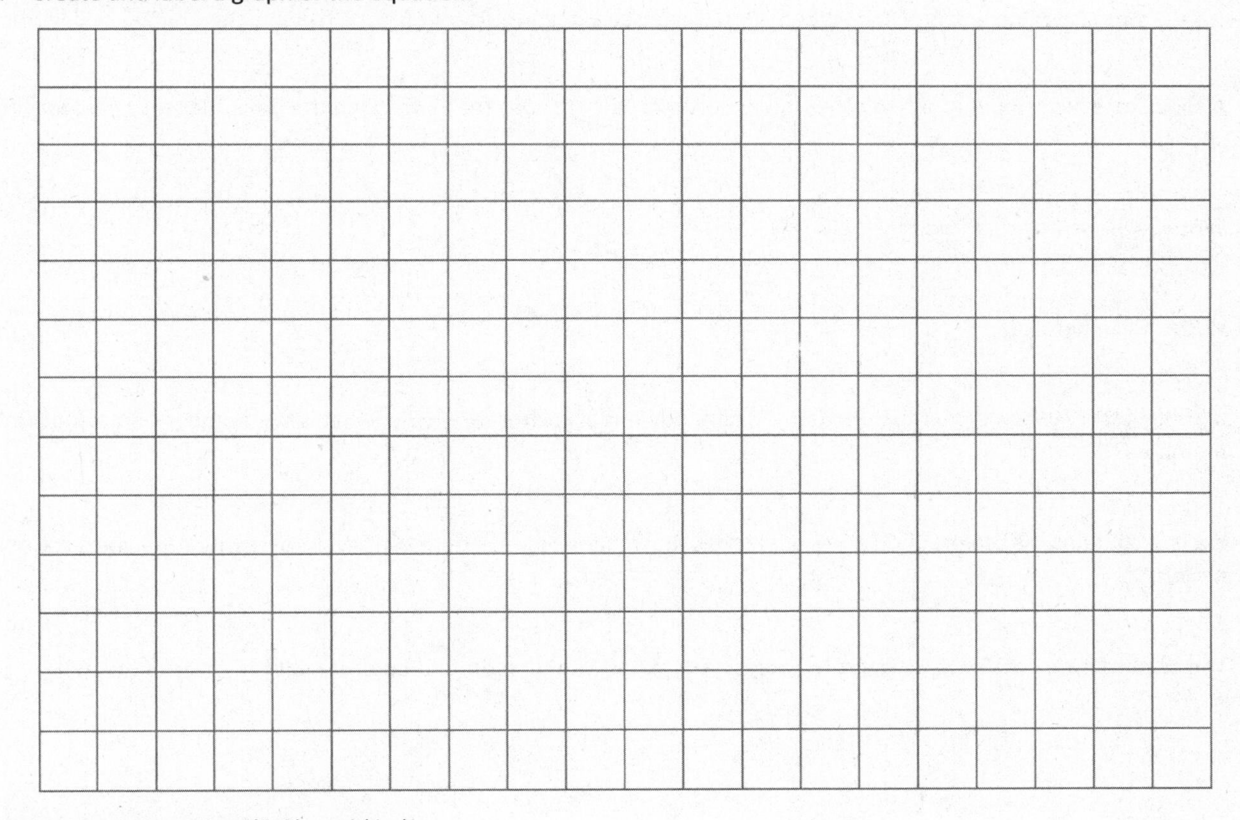

   c. Interpret the points $(0,0)$ and $(1,r)$.

5. At the amusement park, Laura paid $\$6.00$ for a small cotton candy. Her older brother works at the park, and he told her they mark up the cotton candy by 300%. Laura does not think that is mathematically possible. Is it possible, and if so, what is the price of the cotton candy before the markup?

6. A store advertises that customers can take 25% off the original price and then take an extra 10% off. Is this the same as a 35% off discount? Explain.

7. An item that costs $50.00 is marked 20% off. Sales tax for the item is 8%. What is the final price, including tax?
   a. Solve the problem with the discount applied before the sales tax.
   b. Solve the problem with the discount applied after the sales tax.
   c. Compare your answers in parts (a) and (b). Explain.

8. The sale price for a bicycle is $315. The original price was first discounted by 50% and then discounted an additional 10%. Find the original price of the bicycle.

9. A ski shop has a markup rate of 50%. Find the selling price of skis that cost the storeowner $300.

10. A tennis supply store pays a wholesaler $90 for a tennis racquet and sells it for $144. What is the markup rate?

11. A shoe store is selling a pair of shoes for $60 that has been discounted by 25%. What was the original selling price?

12. A shoe store has a markup rate of 75% and is selling a pair of shoes for $133. Find the price the store paid for the shoes.

13. Write $5\frac{1}{4}\%$ as a simple fraction.

14. Write $\frac{3}{8}$ as a percent.

15. If 20% of the 70 faculty members at John F. Kennedy Middle School are male, what is the number of male faculty members?

16. If a bag contains 400 coins, and $33\frac{1}{2}\%$ are nickels, how many nickels are there? What percent of the coins are not nickels?

17. The temperature outside is 60 degrees Fahrenheit. What would be the temperature if it is increased by 20%?

Example 1: How Far Off?

Student	Measurement 1 (in.)	Measurement 2 (in.)
Taylor	$15\frac{2}{8}$	$15\frac{3}{8}$
Connor	$15\frac{4}{8}$	$14\frac{7}{8}$
Jordan	$15\frac{4}{8}$	$14\frac{6}{8}$

Find the absolute error for the following problems. Explain what the absolute error means in context.

a. Taylor's Measurement 1

b. Connor's Measurement 1

c. Jordan's Measurement 2

Example 2: How Right Is Wrong?

a. Find the percent error for Taylor's Measurement 1. What does this mean?

b. From Example 1, part (b), find the percent error for Connor's Measurement 1. What does this mean?

c. From Example 1, part (c), find the percent error for Jordan's Measurement 2. What does it mean?

d. What is the purpose of finding percent error?

Exercises

Calculate the percent error for Problems 1–3. Leave your final answer in fraction form, if necessary.

1. A real estate agent expected 18 people to show up for an open house, but 25 attended.

2. In science class, Mrs. Moore's students were directed to weigh a 300-gram mass on the balance scale. Tina weighed the object and reported 328 grams.

3. Darwin's coach recorded that he had bowled 250 points out of 300 in a bowling tournament. However, the official scoreboard showed that Darwin actually bowled 225 points out of 300.

Example 3: Estimating Percent Error

The attendance at a musical event was counted several times. All counts were between 573 and 589. If the actual attendance number is between 573 and 589, inclusive, what is the most the percent error could be? Explain your answer.

Lesson Summary

- The absolute error is defined as $|a - x|$, where x is the exact value of a quantity and a is an approximate value.
- The percent error is defined as $\frac{|a - x|}{|x|} \times 100\%$.
- The absolute error will tell how big the error is, but the percent error compares the error to the actual value. A good measurement has a small percent error.

Name ____ Date ____________________

1. The veterinarian weighed Oliver's new puppy, Boaz, on a defective scale. He weighed 36 pounds. However, Boaz weighs exactly 34.5 pounds. What is the percent of error in measurement of the defective scale to the nearest tenth?

2. Use the π key on a scientific or graphing calculator to compute the percent of error of the approximation of pi, 3.14, to the value π. Show your steps, and round your answer to the nearest hundredth of a percent.

3. Connor and Angie helped take attendance during their school's practice fire drill. If the actual count was between 77 and 89, inclusive, what is the most the absolute error could be? What is the most the percent error could be? Round your answer to the nearest tenth of a percent.

1. Lincoln High School just installed a new basketball court. The length of the new court is 90 feet, and the width of the court is 49 feet. However, the regulation length and width of a basketball court are 94 ft. and 50 ft.

   a. What is the percent error of the width of the basketball court?

I know the exact value is 50 ft., and the approximate value is 49 ft.

$$\frac{|49-50|}{|50|}\times 100\%$$

I complete the operations within the absolute value before calculating the absolute value, or distance from 0.

$$\frac{|-1|}{|50|}\times 100\%$$

$$\frac{1}{50}\times 100\%$$

$$2\%$$

***The percent error of the width is* 2%.**

   b. The percent error of the area of the basketball court must be less than 5%. If the percent error is larger, the court will have to be re-installed. Does Lincoln High School have to re-install its new basketball court? Why or why not?

Exact Area: $94\text{ ft.}\times 50\text{ ft.} = 4{,}700\text{ ft}^2$

In order to find the percent error of the area, I first calculate the exact area and the approximate area.

Approximate Area: $90\text{ ft.}\times 49\text{ ft.} = 4{,}410\text{ ft}^2$

Percent Error:

$$\frac{|4{,}410-4{,}700|}{|4{,}700|}\times 100\%$$

$$\frac{|-290|}{|4{,}700|}\times 100\%$$

$$\frac{29}{470}\times 100\%$$

$$6.17\%$$

Lincoln High School needs to re-install its basketball court because the area of the new court has more than a* 5% *percent error.

2. Michael volunteered his mom to bring snacks for the seventh grade dance. Michael said that the school is expecting anywhere from 100 to 125 students at the dance. At most, what is the percent error?

The approximate value is 125, and the exact value is 100.

$$\frac{|100-125|}{|125|}\times 100\%$$

$$\frac{|-25|}{|125|}\times 100\%$$

$$\frac{25}{125}\times 100\%$$

$$20\%$$

I know that the percent error is the largest value when the exact value is the smallest value.

The largest percent error is 20%.

3. In the school choir, 76% of the members are female. If there are 350 members in the school choir, how many members are male?

$$100\% - 76\% = 24\%$$

If 76% of the members of female, then the remaining 24% of the members are male.

Let m represent the number of choir members who are male.

$$m = 0.24(350)$$

$$m = 84$$

There are 84 males in the school choir.

1. The odometer in Mr. Washington's car does not work correctly. The odometer recorded 13.2 miles for his last trip to the hardware store, but he knows the distance traveled is 15 miles. What is the percent error? Use a calculator and the percent error formula to help find the answer. Show your steps.

2. The actual length of a soccer field is 500 feet. A measuring instrument shows the length to be 493 feet. The actual width of the field is 250 feet, but the recorded width is 246.5 feet. Answer the following questions based on this information. Round all decimals to the nearest tenth.

   a. Find the percent error for the length of the soccer field.

   b. Find the percent error of the area of the soccer field.

   c. Explain why the values from parts (a) and (b) are different.

250 feet

500 feet

3. Kayla's class went on a field trip to an aquarium. One tank had 30 clown fish. She miscounted the total number of clown fish in the tank and recorded it as 24 fish. What is Kayla's percent error?

4. Sid used geometry software to draw a circle of radius 4 units on a grid. He estimated the area of the circle by counting the squares that were mostly inside the circle and got an answer of 52 square units.

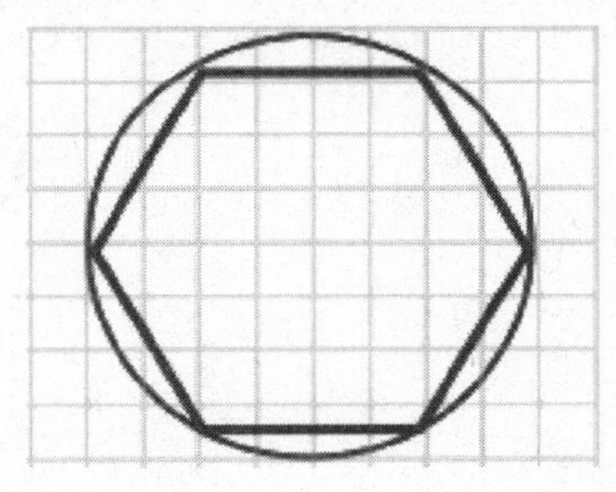

   a. Is his estimate too large or too small?

   b. Find the percent error in Sid's estimation to the nearest hundredth using the π key on your calculator.

5. The exact value for the density of aluminum is 2.699 g/cm^3. Working in the science lab at school, Joseph finds the density of a piece of aluminum to be 2.75 g/cm^3. What is Joseph's percent error? (Round to the nearest hundredth.)

6. The world's largest marathon, The New York City Marathon, is held on the first Sunday in November each year. Between 2 million and 2.5 million spectators will line the streets to cheer on the marathon runners. At most, what is the percent error?

7. A circle is inscribed inside a square, which has a side length of 12.6 cm. Jared estimates the area of the circle to be about 80% of the area of the square and comes up with an estimate of 127 cm^2.

   a. Find the absolute error from Jared's estimate to two decimal places using the π key on your calculator.

   b. Find the percent error of Jared's estimate to two decimal places using the π key on your calculator.

   c. Do you think Jared's estimate was reasonable?

   d. Would this method of computing the area of a circle always be too large?

8. In a school library, 52% of the books are paperback. If there are $2{,}658$ books in the library, how many of them are not paperback to the nearest whole number?

9. Shaniqua has 25% less money than her older sister Jennifer. If Shaniqua has $\$180$, how much money does Jennifer have?

10. An item that was selling for $\$1{,}102$ is reduced to $\$806$. To the nearest whole, what is the percent decrease?

11. If 60 calories from fat is 75% of the total number of calories in a bag of chips, find the total number of calories in the bag of chips.

Example 1

The amount of money Tom has is 75% of Sally's amount of money. After Sally spent $\$120$ and Tom saved all his money, Tom's amount of money is 50% more than Sally's. How much money did each have at the beginning? Use a visual model and a percent line to solve the problem.

Example 2

Erin and Sasha went to a candy shop. Sasha bought 50% more candies than Erin. After Erin bought 8 more candies, Sasha had 20% more. How many candies did Erin and Sasha have at first?

a. Model the situation using a visual model.

b. How many candies did Erin have at first? Explain.

Example 3

Kimberly and Mike have an equal amount of money. After Kimberly spent $\$50$ and Mike spent $\$25$, Mike's money is 50% more than Kimberly's. How much did Kimberly and Mike have at first?

a. Use an equation to solve the problem.

b. Use a visual model to solve the problem.

c. Which method do you prefer and why?

Exercise

Todd has 250% more video games than Jaylon. Todd has 56 video games in his collection. He gives Jaylon 8 of his games. How many video games did Todd and Jaylon have in the beginning? How many do they have now?

Lesson Summary

- To solve a changing percent problem, identify the first whole and then the second whole. To relate the part, whole, and percent, use the formula

$$\text{Quantity} = \text{Percent} \times \text{Whole}.$$

- Models, such as double number lines, can help visually show the change in quantities and percents.

Name ____ Date ____________________

Terrence and Lee were selling magazines for a charity. In the first week, Terrance sold 30% more than Lee. In the second week, Terrance sold 8 magazines, but Lee did not sell any. If Terrance sold 50% more than Lee by the end of the second week, how many magazines did Lee sell?

Choose any model to solve the problem. Show your work to justify your answer.

Use models to solve each problem.

1. The number of video games Scott has is 125% of the number of Doug's video games. However, Scott's mother forced him to donate 8 video games. Now, Scott and Doug have the same number of video games. How many video games did each boy start with?

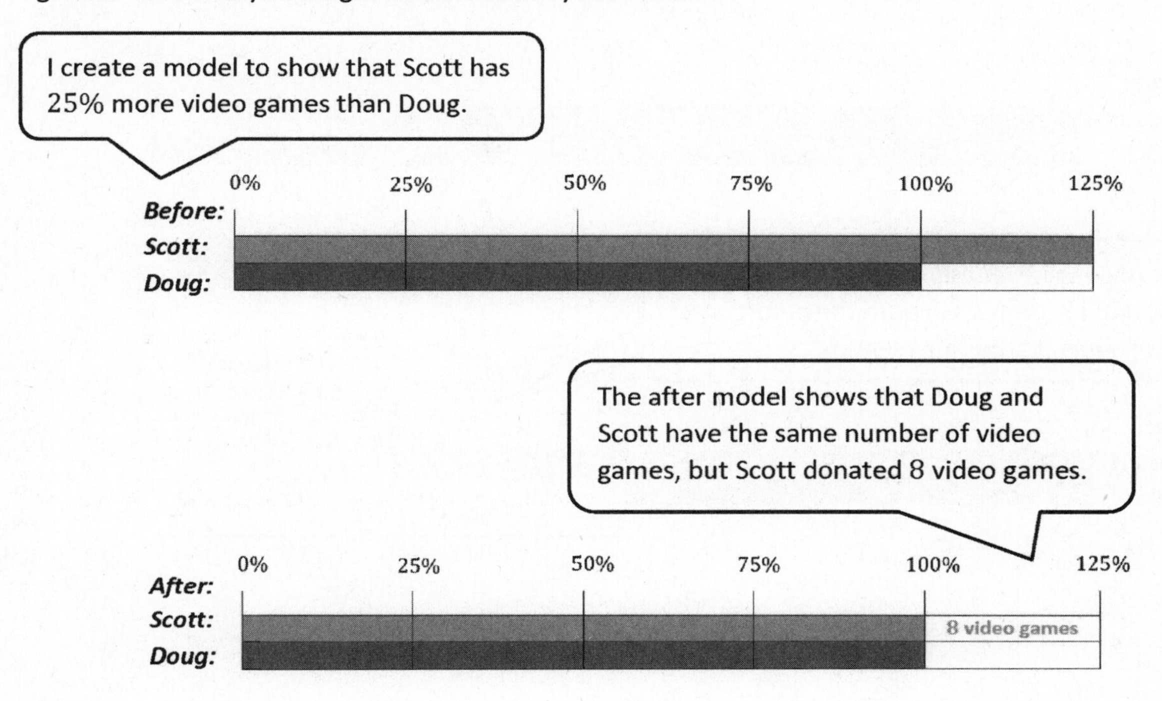

The models show that each bar represents 8 video games.

Scott started with 40 video games because his original tape diagram has 5 bars that each represent 8 video games.

Doug started with 32 video games because his original tape diagram has 4 bars that, again, each represent 8 video games.

2. Genesis and Kyle went to the store to buy school supplies. Genesis bought 75% as many pencils as Kyle. Kyle ended up giving 10 of his pencils to his friend, and now the number of pencils Genesis has is 50% more than Kyle. How many pencils did each person have at the beginning?

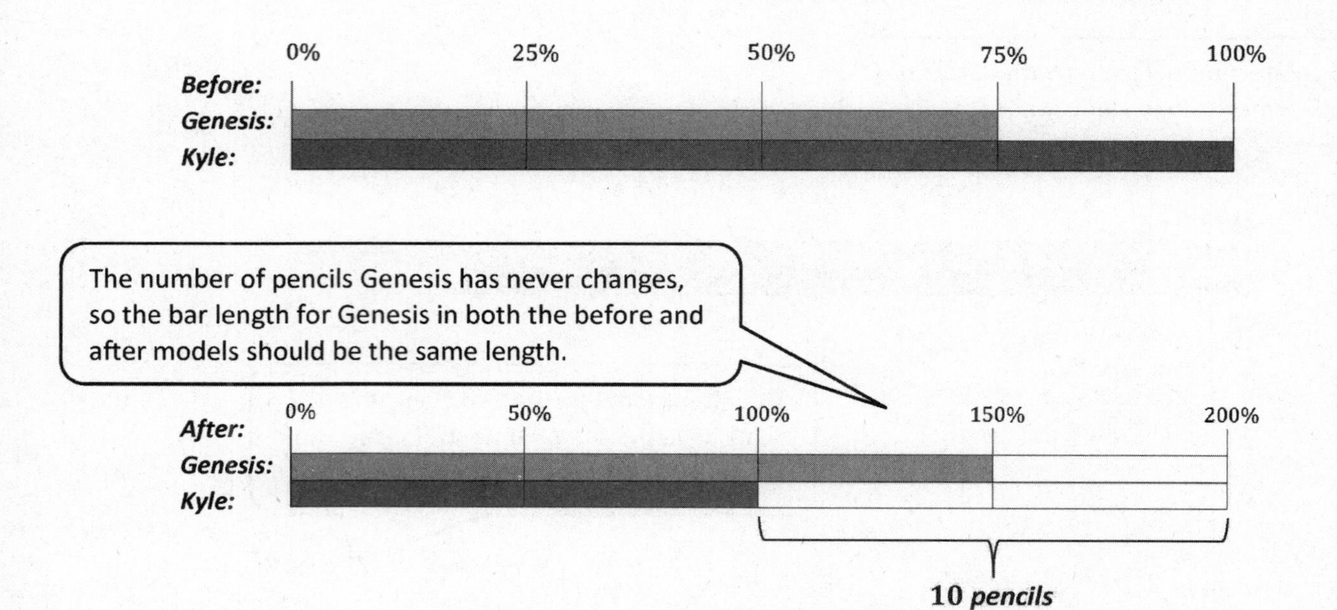

The models show that each bar represents 5 pencils because two bars represent 10 pencils.

Genesis started with 15 pencils because her tape diagram has 3 bars where each represents 5 pencils.

Kyle started with 20 pencils because his tape diagram has 4 bars where each represents 5 pencils.

3. Janaye and Nevaeh compared their money and noticed Janaye had 50% more money than Nevaeh. After Nevaeh earned an extra \$20, Janaye had 20% more money. How much more money did Janaye have than Nevaeh at first?

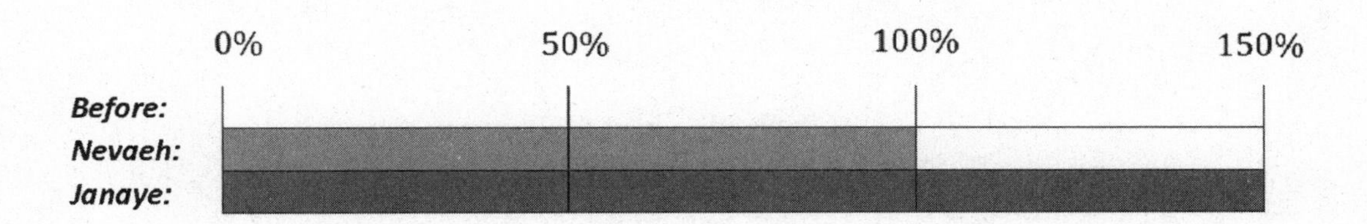

Each model has a different size bars; in order to solve the problem the bars must all be the same size.

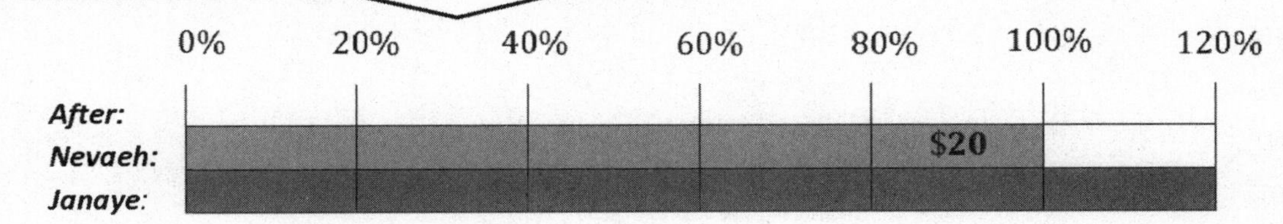

The bars in the before model need to be split in half in order to be the same size as the bars in the after model. This is shown below.

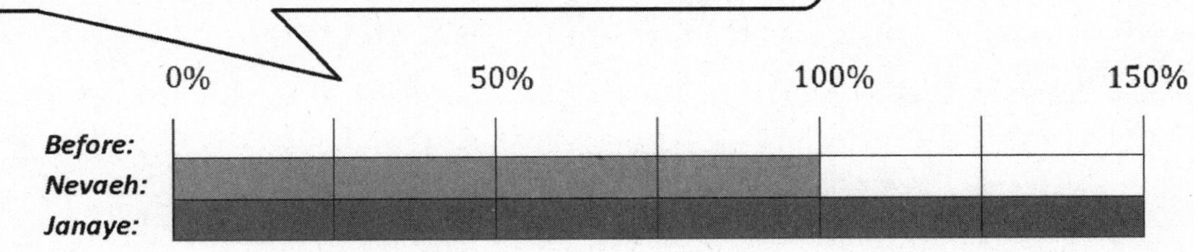

***Now that the bars on both models are the same size, I know each bar represents* \$20.**

Therefore, Nevaeh had* \$80 *at the beginning because her tape diagram has* 4 *bars where each represents* \$20, *and Janaye had* \$120 *at first.

In order to answer the question, I have to determine how much each girl had at the beginning.

$\$120 - \$80 = \$40$

This means that Janaye had* \$40 *more than Nevaeh at the start.

1. Solve each problem using an equation.
   a. What is 150% of 625?
   b. 90 is 40% of what number?
   c. What percent of 520 is 40? Round to the nearest hundredth of a percent.

2. The actual length of a machine is 12.25 cm. The measured length is 12.2 cm. Round the answer to part (b) to the nearest hundredth of a percent.
   a. Find the absolute error.
   b. Find the percent error.

3. A rowing club has 600 members. 60% of them are women. After 200 new members joined the club, the percentage of women was reduced to 50%. How many of the new members are women?

4. 40% of the marbles in a bag are yellow. The rest are orange and green. The ratio of the number of orange to the number of green is 4 : 5. If there are 30 green marbles, how many yellow marbles are there? Use a visual model to show your answer.

5. Susan has 50% more books than Michael. Michael has 40 books. If Michael buys 8 more books, will Susan have more or less books than Michael? What percent more or less will Susan's books be? Use any method to solve the problem.

6. Harry's amount of money is 75% of Kayla's amount of money. After Harry earned $30 and Kayla earned 25% more of her money, Harry's amount of money is 80% of Kayla's money. How much money did each have at the beginning? Use a visual model to solve the problem.

Number Correct: ______

Fractional Percents—Round 1

Directions: Find the part that corresponds with each percent.

1.	1% of 100	
2.	1% of 200	
3.	1% of 400	
4.	1% of 800	
5.	1% of 1,600	
6.	1% of 3,200	
7.	1% of 5,000	
8.	1% of 10,000	
9.	1% of 20,000	
10.	1% of 40,000	
11.	1% of 80,000	
12.	$\frac{1}{2}$% of 100	
13.	$\frac{1}{2}$% of 200	
14.	$\frac{1}{2}$% of 400	
15.	$\frac{1}{2}$% of 800	
16.	$\frac{1}{2}$% of 1,600	
17.	$\frac{1}{2}$% of 3,200	
18.	$\frac{1}{2}$% of 5,000	
19.	$\frac{1}{2}$% of 10,000	
20.	$\frac{1}{2}$% of 20,000	
21.	$\frac{1}{2}$% of 40,000	
22.	$\frac{1}{2}$% of 80,000	

23.	$\frac{1}{4}$% of 100	
24.	$\frac{1}{4}$% of 200	
25.	$\frac{1}{4}$% of 400	
26.	$\frac{1}{4}$% of 800	
27.	$\frac{1}{4}$% of 1,600	
28.	$\frac{1}{4}$% of 3,200	
29.	$\frac{1}{4}$% of 5,000	
30.	$\frac{1}{4}$% of 10,000	
31.	$\frac{1}{4}$% of 20,000	
32.	$\frac{1}{4}$% of 40,000	
33.	$\frac{1}{4}$% of 80,000	
34.	1% of 1,000	
35.	$\frac{1}{2}$% of 1,000	
36.	$\frac{1}{4}$% of 1,000	
37.	1% of 4,000	
38.	$\frac{1}{2}$% of 4,000	
39.	$\frac{1}{4}$% of 4,000	
40.	1% of 2,000	
41.	$\frac{1}{2}$% of 2,000	
42.	$\frac{1}{4}$% of 2,000	
43.	$\frac{1}{2}$% of 6,000	
44.	$\frac{1}{4}$% of 6,000	

Number Correct: ______

Fractional Percents—Round 2

Improvement: ______

Directions: Find the part that corresponds with each percent.

1.	10% of 30		23.	$10\frac{1}{2}$% of 100	
2.	10% of 60		24.	$10\frac{1}{2}$% of 200	
3.	10% of 90		25.	$10\frac{1}{2}$% of 400	
4.	10% of 120		26.	$10\frac{1}{2}$% of 800	
5.	10% of 150		27.	$10\frac{1}{2}$% of 1,600	
6.	10% of 180		28.	$10\frac{1}{2}$% of 3,200	
7.	10% of 210		29.	$10\frac{1}{2}$% of 6,400	
8.	20% of 30		30.	$10\frac{1}{4}$% of 400	
9.	20% of 60		31.	$10\frac{1}{4}$% of 800	
10.	20% of 90		32.	$10\frac{1}{4}$% of 1,600	
11.	20% of 120		33.	$10\frac{1}{4}$% of 3,200	
12.	5% of 50		34.	10% of 1,000	
13.	5% of 100		35.	$10\frac{1}{2}$% of 1,000	
14.	5% of 200		36.	$10\frac{1}{4}$% of 1,000	
15.	5% of 400		37.	10% of 2,000	
16.	5% of 800		38.	$10\frac{1}{2}$% of 2,000	
17.	5% of 1,600		39.	$10\frac{1}{4}$% of 2,000	
18.	5% of 3,200		40.	10% of 4,000	
19.	5% of 6,400		41.	$10\frac{1}{2}$% of 4,000	
20.	5% of 600		42.	$10\frac{1}{4}$% of 4,000	
21.	10% of 600		43.	10% of 5,000	
22.	20% of 600		44.	$10\frac{1}{2}$% of 5,000	

To find the simple interest, use the following formula:

$$\text{Interest} = \text{Principal} \times \text{Rate} \times \text{Time}$$
$$I = P \times r \times t$$
$$I = Prt$$

- r is the percent of the principal that is paid over a period of time (usually per year).
- t is the time.
- r and t must be compatible. For example, if r is an annual interst rate, then t must be written in years.

Example 1: Can Money Grow? A Look at Simple Interest

Larry invests $\$100$ in a savings plan. The plan pays $4\frac{1}{2}\%$ interest each year on his $\$100$ account balance.

a. How much money will Larry earn in interest after 3 years? After 5 years?

b. How can you find the balance of Larry's account at the end of 5 years?

Exercise 1

Find the balance of a savings account at the end of 10 years if the interest earned each year is 7.5%. The principal is $\$500$.

Example 2: Time Other Than One Year

A $\$1,000$ savings bond earns simple interest at the rate of 3% each year. The interest is paid at the end of every month. How much interest will the bond have earned after 3 months?

Example 3: Solving for P, r, or t

Mrs. Williams wants to know how long it will take an investment of $\$450$ to earn $\$200$ in interest if the yearly interest rate is 6.5%, paid at the end of each year.

Exercise 2

Write an equation to find the amount of simple interest, A, earned on a $\$600$ investment after $1\frac{1}{2}$ years if the semiannual (6-month) interest rate is 2%.

Exercise 3

A $\$1{,}500$ loan has an annual interest rate of $4\frac{1}{4}\%$ on the amount borrowed. How much time has elapsed if the interest is now $\$127.50$?

Lesson Summary

- Interest earned over time can be represented by a proportional relationship between time, in years, and interest.
- The simple interest formula is

$$\text{Interest} = \text{Principal} \times \text{Rate} \times \text{Time}$$
$$I = P \times r \times t$$
$$I = Prt$$

  r is the percent of the principal that is paid over a period of time (usually per year).
  t is the time.

- The rate, r, and time, t, must be compatible. If r is the annual interest rate, then t must be written in years.

Name ______________________________ Date ______________

1. Erica's parents gave her $\$500$ for her high school graduation. She put the money into a savings account that earned 7.5% annual interest. She left the money in the account for nine months before she withdrew it. How much interest did the account earn if interest is paid monthly?

2. If she would have left the money in the account for another nine months before withdrawing, how much interest would the account have earned?

3. About how many years and months would she have to leave the money in the account if she wants to reach her goal of saving $\$750$?

1. Joy borrowed \$2,000 from the bank and agreed to pay an annual interest rate of 6% for 24 months. What is the amount of interest she will pay on this loan?

$$I = Prt$$

$$I = (2{,}000)(6\%)\left(\frac{24}{12}\right)$$

> I have annual interest, which means the time must be represented in years.

$$I = (2{,}000)(0.06)(2)$$

$$I = 240$$

Joy will pay* $\$240$ *in interest.

2. Xavier opened a new savings account by putting \$300 into the account. At the end of the year, the savings account pays an annual interest of 5.25% on the amount he put into the account.

   a. How much will Xavier earn if he leaves his money in the savings account for 15 years?

> I know the principal amount is \$300, the annual interest rate is 5.25%, and the time is 15 years.

$$I = Prt$$

$$I = (300)(5.25\%)(15)$$

> Before multiplying, I must convert the percent to a decimal.

$$I = (300)(0.0525)(15)$$

$$I = 236.25$$

Xavier will earn* $\$236.25$ *in interest.

   b. If Xavier does not add any money to his account, how much money will Xavier have in his account after the 15 years?

$$\$300 + \$236.25 = \$536.25$$

After* 15 *years, Xavier will have* $\$536.25$ *in his account.

3. Mr. Brown had to take out a loan to help pay rent. He borrowed \$800 and agreed to pay an annual interest rate of 6%. If Mr. Brown was able to pay the loan back in just 6 months, how much interest did Mr. Brown pay?

$\mathbf{12}$ ***months*** $\mathbf{= 1}$ ***year***

If we divide both sides by $\mathbf{2}$***, we find that*** $\mathbf{6}$ ***months is equal to*** $\frac{1}{2}$ ***year.***

$I = Prt$

$I = 800(0.06)\left(\frac{1}{2}\right)$

$I = 24$

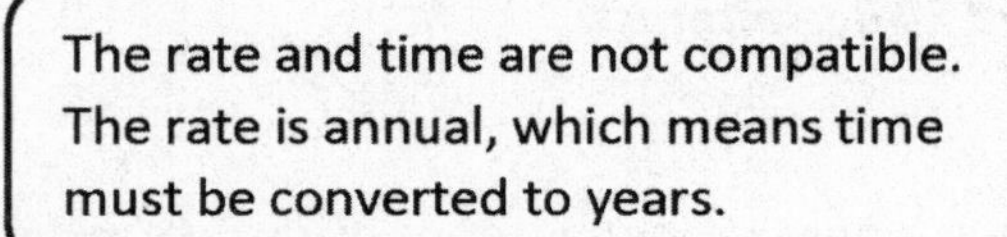

Mr. Brown paid $\mathbf{\$24}$ ***in interest.***

4. Stefani received a loan for \$2,000 and now has acquired \$720 in interest. If she pays an annual interest of 4.5% on the amount borrowed, how much time has elapsed since Stefani received the loan?

Let t represent the time, in years, that has elapsed since Stefani received the loan.

$$I = Prt$$

$$720 = 2{,}000(0.045)t$$

$$720 = 90t$$

$$\left(\frac{1}{90}\right)(720) = \left(\frac{1}{90}\right)90t$$

$$8 = t$$

This time, I know the interest and need to calculate the time (in years).

Therefore, Stefani received the loan $\mathbf{8}$ ***years ago.***

1. Enrique takes out a student loan to pay for his college tuition this year. Find the interest on the loan if he borrowed $\$2,500$ at an annual interest rate of 6% for 15 years.

2. Your family plans to start a small business in your neighborhood. Your father borrows $\$10,000$ from the bank at an annual interest rate of 8% rate for 36 months. What is the amount of interest he will pay on this loan?

3. Mr. Rodriguez invests $\$2,000$ in a savings plan. The savings account pays an annual interest rate of 5.75% on the amount he put in at the end of each year.
   a. How much will Mr. Rodriguez earn if he leaves his money in the savings plan for 10 years?
   b. How much money will be in his savings plan at the end of 10 years?
   c. Create (and label) a graph in the coordinate plane to show the relationship between time and the amount of interest earned for 10 years. Is the relationship proportional? Why or why not? If so, what is the constant of proportionality?
   d. Explain what the points $(0,0)$ and $(1,115)$ mean on the graph.
   e. Using the graph, find the balance of the savings plan at the end of seven years.
   f. After how many years will Mr. Rodriguez have increased his original investment by more than 50%? Show your work to support your answer.

Challenge Problem

4. George went on a game show and won $\$60,000$. He wanted to invest it and found two funds that he liked. Fund 250 earns 15% interest annually, and Fund 100 earns 8% interest annually. George does not want to earn more than $\$7,500$ in interest income this year. He made the table below to show how he could invest the money.

	I	P	r	t
Fund 100		x	0.08	1
Fund 250		$60,000 - x$	0.15	1
Total	$7,500$	$60,000$		

   a. Explain what value x is in this situation.
   b. Explain what the expression $60,000 - x$ represents in this situation.
   c. Using the simple interest formula, complete the table for the amount of interest earned.
   d. Write an inequality to show the total amount of interest earned from both funds.
   e. Use algebraic properties to solve for x and the principal, in dollars, George could invest in Fund 100. Show your work.
   f. Use your answer from part (e) to determine how much George could invest in Fund 250.
   g. Using your answers to parts (e) and (f), how much interest would George earn from each fund?

Opening Exercise: Tax, Commission, Gratuity, and Fees

How are each of the following percent applications different, and how are they the same? Solve each problem, and then compare your solution process for each problem.

a. Silvio earns 10% for each car sale he makes while working at a used car dealership. If he sells a used car for $\$2{,}000$, what is his commission?

b. Tu's family stayed at a hotel for 10 nights on their vacation. The hotel charged a 10% room tax, per night. How much did they pay in room taxes if the room cost $\$200$ per night?

c. Eric bought a new computer and printer online. He had to pay 10% in shipping fees. The items totaled $\$2{,}000$. How much did the shipping cost?

d. Selena had her wedding rehearsal dinner at a restaurant. The restaurant's policy is that gratuity is included in the bill for large parties. Her father said the food and service were exceptional, so he wanted to leave an extra 10% tip on the total amount of the bill. If the dinner bill totaled $\$2{,}000$, how much money did her father leave as the extra tip?

Exercises

Show all work; a calculator may be used for calculations.

The school board has approved the addition of a new sports team at your school.

1. The district ordered 30 team uniforms and received a bill for $\$2{,}992.50$. The total included a 5% discount.

   a. The school needs to place another order for two more uniforms. The company said the discount will not apply because the discount only applies to orders of $\$1{,}000$ or more. How much will the two uniforms cost?

   b. The school district does not have to pay the 8% sales tax on the $\$2{,}992.50$ purchase. Estimate the amount of sales tax the district saved on the $\$2{,}992.50$ purchase. Explain how you arrived at your estimate.

   c. A student who loses a uniform must pay a fee equal to 75% of the school's cost of the uniform. For a uniform that cost the school $\$105$, will the student owe more or less than $\$75$ for the lost uniform? Explain how to use mental math to determine the answer.

   d. Write an equation to represent the proportional relationship between the school's cost of a uniform and the amount a student must pay for a lost uniform. Use u to represent the uniform cost and s to represent the amount a student must pay for a lost uniform. What is the constant of proportionality?

2. A taxpayer claims the new sports team caused his school taxes to increase by 2%.

   a. Write an equation to show the relationship between the school taxes before and after a 2% increase. Use b to represent the dollar amount of school tax before the 2% increase and t to represent the dollar amount of school tax after the 2% increase.

   b. Use your equation to complete the table below, listing at least 5 pairs of values.

b	t
1,000	
2,000	
	3,060
	6,120

   c. On graph paper, graph the relationship modeled by the equation in part (a). Be sure to label the axes and scale.

   d. Is the relationship proportional? Explain how you know.

   e. What is the constant of proportionality? What does it mean in the context of the situation?

   f. If a taxpayers' school taxes rose from $4,000 to $4,020, was there a 2% increase? Justify your answer using your graph, table, or equation.

3. The sports booster club sold candles as a fundraiser to support the new team. The club earns a commission on its candle sales (which means it receives a certain percentage of the total dollar amount sold). If the club gets to keep 30% of the money from the candle sales, what would the club's total sales have to be in order to make at least $\$500$?

4. Christian's mom works at the concession stand during sporting events. She told him they buy candy bars for $\$0.75$ each and mark them up 40% to sell at the concession stand. What is the amount of the markup? How much does the concession stand charge for each candy bar?

With your group, brainstorm solutions to the problems below. Prepare a poster that shows your solutions and math work. A calculator may be used for calculations.

5. For the next school year, the new soccer team will need to come up with $\$600$.

   a. Suppose the team earns $\$500$ from the fundraiser at the start of the current school year, and the money is placed for one calendar year in a savings account earning 0.5% simple interest annually. How much money will the team still need to raise to meet next year's expenses?

   b. Jeff is a member of the new sports team. His dad owns a bakery. To help raise money for the team, Jeff's dad agrees to provide the team with cookies to sell at the concession stand for next year's opening game. The team must pay back the bakery $\$0.25$ for each cookie it sells. The concession stand usually sells about 60 to 80 baked goods per game. Using your answer from part (a), determine a percent markup for the cookies the team plans to sell at next year's opening game. Justify your answer.

c. Suppose the team ends up selling 78 cookies at next year's opening game. Find the percent error in the number of cookies that you estimated would be sold in your solution to part (b).

Percent Error $= \frac{|a - x|}{|x|} \cdot 100\%$, where x is the exact value and a is the approximate value.

Lesson Summary

- There are many real-world problems that involve percents. For example, gratuity (tip), commission, fees, and taxes are applications found daily in the real world. They each increase the total, so all questions like these reflect a percent increase. Likewise, markdowns and discounts decrease the total, so they reflect a percent decrease.
- Regardless of the application, the percent relationship can be represented as

$$\text{Quantity(Part)} = \text{Percent(\%)} \times \text{Whole}$$

Name ____ Date ____________________

Lee sells electronics. He earns a 5% commission on each sale he makes.

a. Write an equation that shows the proportional relationship between the dollar amount of electronics Lee sells, d, and the amount of money he makes in commission, c.

b. Express the constant of proportionality as a decimal.

c. Explain what the constant of proportionality means in the context of this situation.

d. If Lee wants to make $100 in commission, what is the dollar amount of electronics he must sell?

1. In order for Josue to pay all of his bills, he needs to make $1,900 a month. Josue earns $10 an hour, plus 8% commission on his insurance sales. If Josue worked 160 hours each month, what is the least amount in insurance Josue would have to sell in order to have enough money to pay his bills?

Let s represent the amount, in dollars, sold in insurance.

The 1,900 represents the amount Josue needs to earn each month.

These factors will tell me how much Josue gets paid each month, without commission.

These factors will tell me the amount of commission Josue will earn.

$$1{,}900 = 160(10) + 8\%(s)$$
$$1{,}900 = 1{,}600 + 0.08s$$
$$1{,}900 - 1{,}600 = 1{,}600 + 0.08s - 1{,}600$$
$$300 = 0.08s$$
$$\left(\frac{1}{0.08}\right)(300) = \left(\frac{1}{0.08}\right)(0.08)s$$
$$3{,}750 = s$$

Therefore, Josue would have to sell $3,750 worth of insurance in order to pay his monthly bills.

2. A dealership sells a car to an average of 12% of the daily customers.

   a. Write an equation that shows the proportional relationship between the number of customers who go to the dealership, c, and the number of customers who actually buy a car, b.

   $b = 0.12c$

   I will multiply the percent of customers who buy a car by the number of daily customers in order to determine the number of customers who buy a car.

b. Use your equation to complete the table. List 5 possible values for b and c.

c	b
50	6
100	12
150	18
200	24
250	30

I can choose any numbers for c because this is the independent variable.

In order to determine my b values, I have to use my equation. For example, $b = 0.12(50)$. After multiplying, I find $b = 6$.

c. Identify the constant of proportionality, and explain what it means in the context of the situation.

The constant of proportionality is 0.12, or 12%. On average, for every 100 customers who go to the car dealership, 12 will buy a car.

1. A school district's property tax rate rises from 2.5% to 2.7% to cover a $\$300,000$ budget deficit (shortage of money). What is the value of the property in the school district to the nearest dollar? (Note: Property is assessed at 100% of its value.)

2. Jake's older brother Sam has a choice of two summer jobs. He can either work at an electronics store or at the school's bus garage. The electronics store would pay him to work 15 hours per week. He would make $\$8$ per hour plus a 2% commission on his electronics sales. At the school's bus garage, Sam could earn $\$300$ per week working 15 hours cleaning buses. Sam wants to take the job that pays him the most. How much in electronics would Sam have to sell for the job at the electronics store to be the better choice for his summer job?

3. Sarah lost her science book. Her school charges a lost book fee equal to 75% of the cost of the book. Sarah received a notice stating she owed the school $\$60$ for the lost book.
   a. Write an equation to represent the proportional relationship between the school's cost for the book and the amount a student must pay for a lost book. Let B represent the school's cost of the book in dollars and N represent the student's cost in dollars.
   b. What is the constant or proportionality? What does it mean in the context of this situation?
   c. How much did the school pay for the book?

4. In the month of May, a certain middle school has an average daily absentee rate of 8% each school day. The absentee rate is the percent of students who are absent from school each day.
   a. Write an equation that shows the proportional relationship between the number of students enrolled in the middle school and the average number of students absent each day during the month of May. Let s represent the number of students enrolled in school, and let a represent the average number of students absent each day in May.
   b. Use your equation to complete the table. List 5 possible values for s and a.

s	a

   c. Identify the constant of proportionality, and explain what it means in the context of this situation.
   d. Based on the absentee rate, determine the number of students absent on average from school during the month of May if there are 350 students enrolled in the middle school.

5. The equation shown in the box below could relate to many different percent problems. Put an X next to each problem that could be represented by this equation. For any problem that does not match this equation, explain why it does not. $\boxed{\text{Quantity} = 1.05 \cdot \text{Whole}}$

_______ Find the amount of an investment after 1 year with 0.5% interest paid annually.

_______ Write an equation to show the amount paid for an item including tax, if the tax rate is 5%.

_______ A proportional relationship has a constant of proportionality equal to 105%.

____

Whole	0	100	200	300	400	500
Quantity	0	105	210	315	420	525

_______ Mr. Hendrickson sells cars and earns a 5% commission on every car he sells. Write an equation to show the relationship between the price of a car Mr. Hendrickson sold and the amount of commission he earns.

Opening

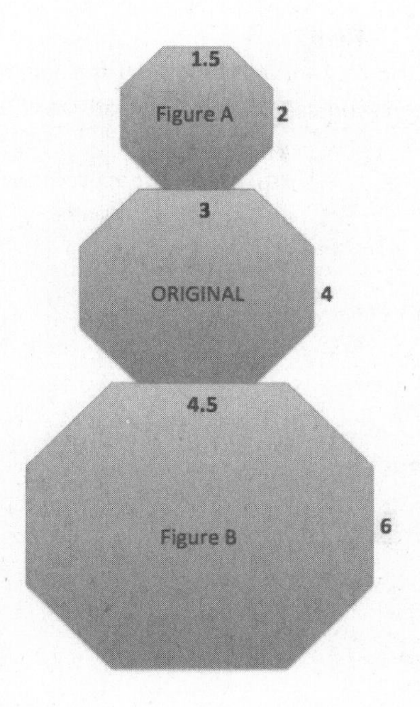

Compare the corresponding lengths of Figure A to the original octagon in the middle. This is an example of a particular type of *scale drawing* called a

________________________. Explain why it is called that.

Compare the corresponding lengths of Figure B to the original octagon in the middle. This is an example of a particular type of *scale drawing* called an

__________________________. Explain why it is called that.

The *scale factor* is the quotient of any length in the scale drawing to its corresponding length in the actual drawing.

Use what you recall from Module 1 to determine the scale factors between the original figure and Figure A and the original figure and Figure B.

Use the diagram to complete the chart below to determine the horizontal and vertical scale factors. Write answers as a percent and as a concluding statement using the previously learned reduction and enlargement vocabulary.

	Horizontal Measurement in Scale Drawing	Vertical Measurement in Scale Drawing	Concluding Statement
Figure A			
Figure B			

Example 1

Create a snowman on the accompanying grid. Use the octagon given as the middle of the snowman with the following conditions:

a. Calculate the width, neck, and height, in units, for the figure to the right.

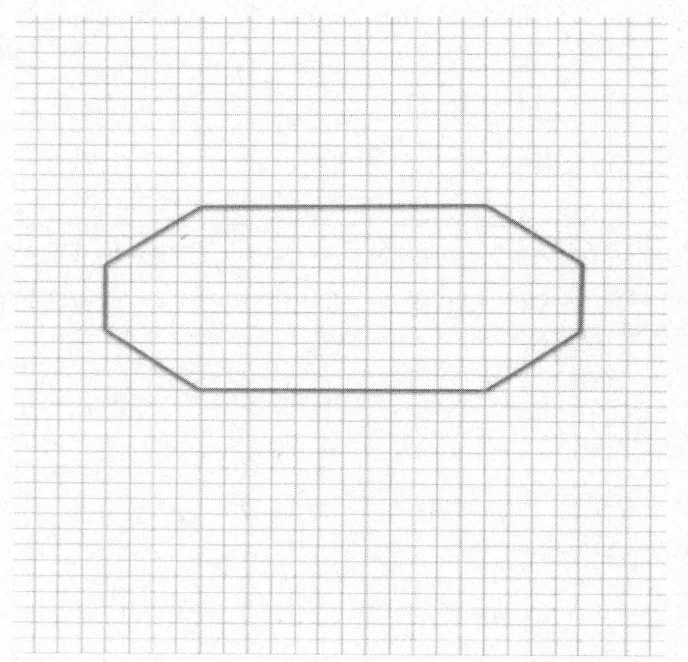

b. To create the head of the snowman, make a scale drawing of the middle of the snowman with a scale factor of 75%. Calculate the new lengths, in units, for the width, neck, and height.

c. To create the bottom of the snowman, make a scale drawing of the middle of the snowman with a scale factor of 125%. Calculate the new lengths, in units, for the width, waist, and height.

d. Is the head a reduction or enlargement of the middle?

e. Is the bottom a reduction or enlargement of the middle?

f. What is the significance of the scale factor as it relates to 100%? What happens when such scale factors are applied?

g. Use the dimensions you calculated in parts (b) and (c) to draw the complete snowman.

Example 2

Create a scale drawing of the arrow below using a scale factor of 150%.

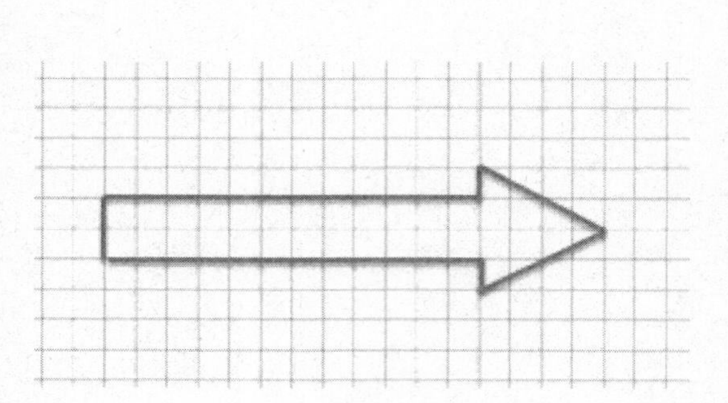

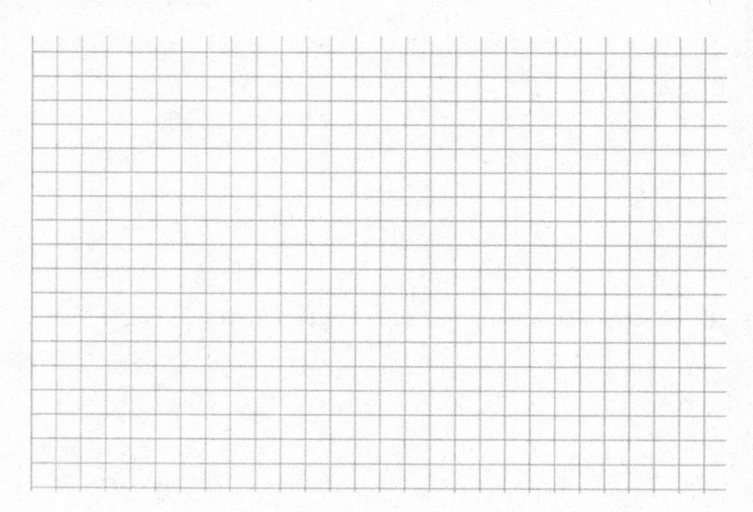

Example 3: Scale Drawings Where the Horizontal and Vertical Scale Factors Are Different

Sometimes it is helpful to make a scale drawing where the horizontal and vertical scale factors are different, such as when creating diagrams in the field of engineering. Having differing scale factors may distort some drawings. For example, when you are working with a very large horizontal scale, you sometimes must exaggerate the vertical scale in order to make it readable. This can be accomplished by creating a drawing with two scales. Unlike the scale drawings with just one scale factor, these types of scale drawings may look distorted. Next to the drawing below is a scale drawing with a horizontal scale factor of 50% and vertical scale factor of 25% (given in two steps). Explain how each drawing is created.

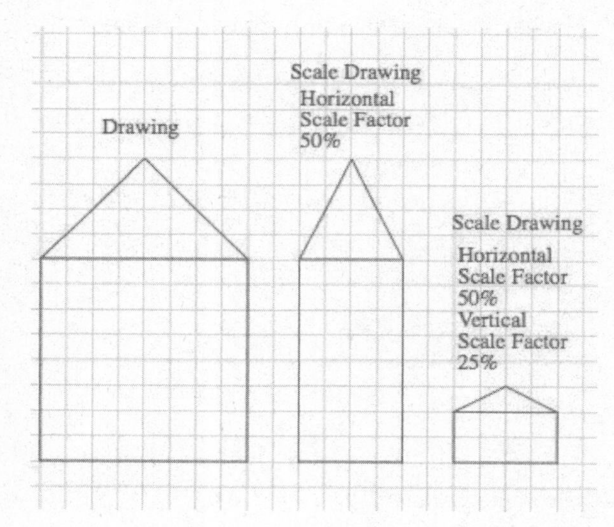

Exercise 1

Create a scale drawing of the following drawing using a horizontal scale factor of $183\frac{1}{3}\%$ and a vertical scale factor of 25%.

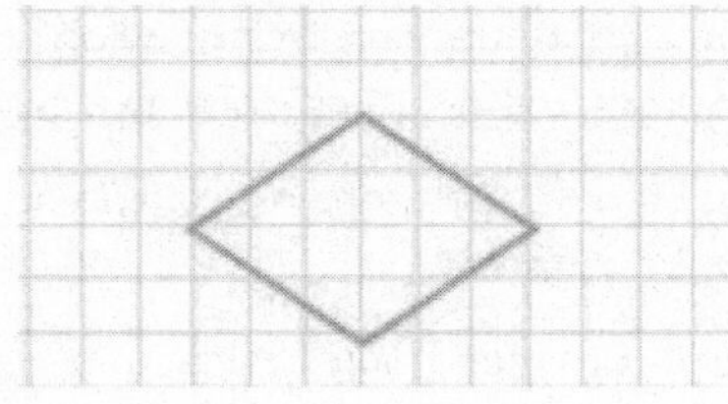

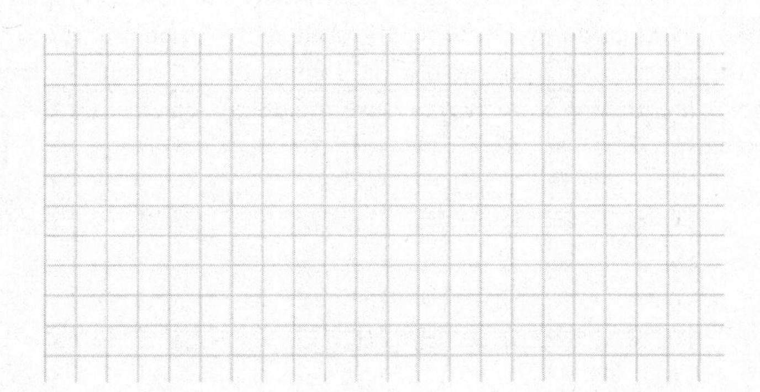

Exercise 2

Chris is building a rectangular pen for his dog. The dimensions are 12 units long and 5 units wide.

Chris is building a second pen that is 60% the length of the original and 125% the width of the original. Write equations to determine the length and width of the second pen.

Lesson Summary

The scale factor is the number that determines whether the new drawing is an enlargement or a reduction of the original. If the scale factor is greater than 100%, then the resulting drawing is an enlargement of the original drawing. If the scale factor is less than 100%, then the resulting drawing is a reduction of the original drawing.

When a scale factor is mentioned, assume that it refers to both vertical and horizontal factors. It is noted if the horizontal and vertical factors are intended to be different.

To create a scale drawing with both the same vertical and horizontal factors, determine the horizontal and vertical distances of the original drawing. Using the given scale factor, determine the new corresponding lengths in the scale drawing by writing a numerical equation that requires the scale factor to be multiplied by the original length. Draw new segments based on the calculations from the original segments. If the scale factors are different, determine the new corresponding lengths the same way but use the unique given scale factor for each of the horizontal length and vertical length.

Name ____________________ Date ____________

1. Create a scale drawing of the picture below using a scale factor of 60%. Write three equations that show how you determined the lengths of three different parts of the resulting picture.

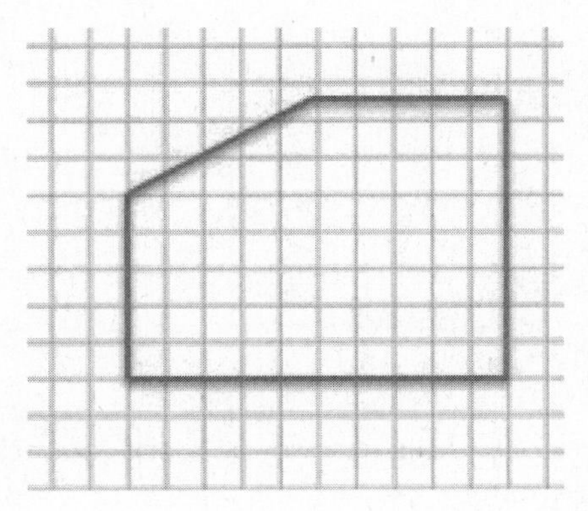

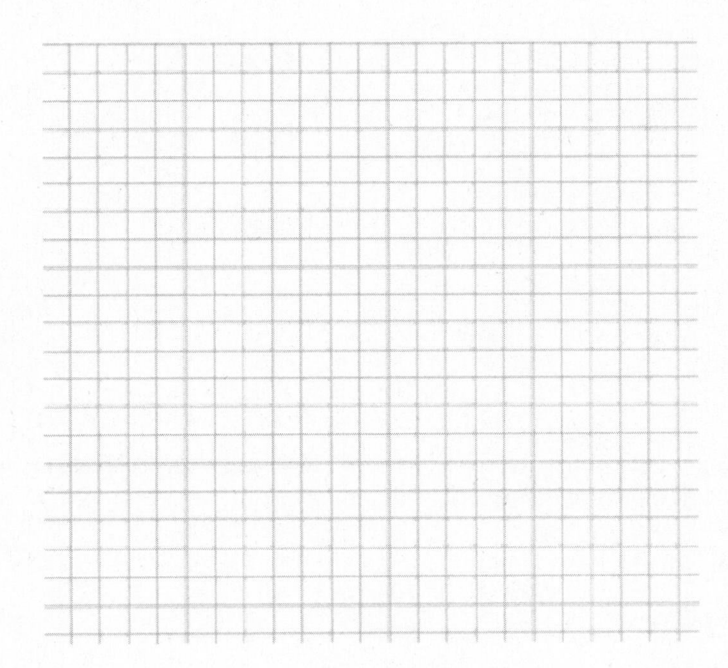

2. Sue wants to make two picture frames with lengths and widths that are proportional to the ones given below. Note: The illustration shown below is not drawn to scale.

a. Sketch a scale drawing using a horizontal scale factor of 50% and a vertical scale factor of 75%. Determine the dimensions of the new picture frame.

b. Sketch a scale drawing using a horizontal scale factor of 125% and a vertical scale factor of 140%. Determine the dimensions of the new picture frame.

The scale factor is less than 100%, which means the scale drawing will be smaller than the original diagram.

1. Use the diagram below to create a scale drawing using a scale factor of 80%. Write numerical equations to find the horizontal and vertical distances in the scale drawing.

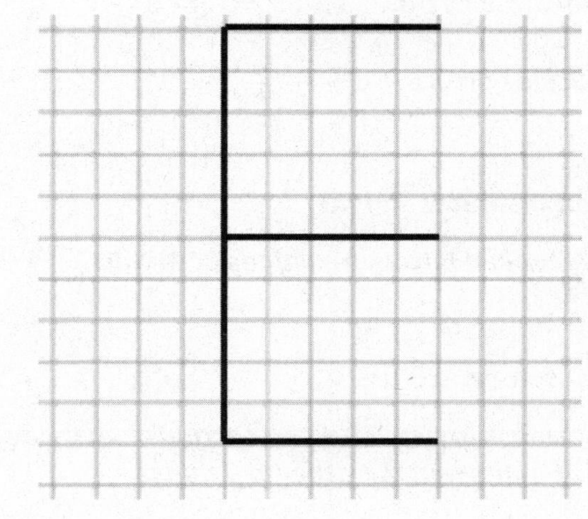

Scale Factor: $80\% = \frac{80}{100} = \frac{4}{5}$

Horizontal Distance: **5 *units***

Vertical Distance: **10 *units (broken into two equal sections of* 5 *units).***

To determine the dimensions of the scale drawing, I multiply each dimension by the scale factor of 80%, or $\frac{4}{5}$.

Scale Drawing:

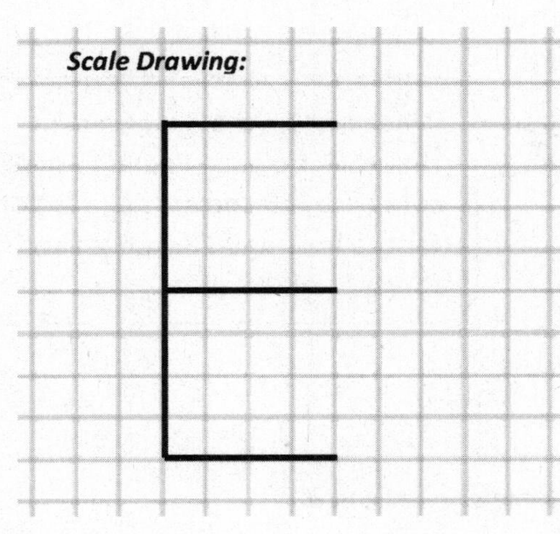

Horizontal Distance of Scale Drawing:

$(5 \text{ units})\left(\frac{4}{5}\right) = 4 \text{ units}$

Vertical Distance of Scale Drawing:

$(10 \text{ units})\left(\frac{4}{5}\right) = 8 \text{ units}$ ***(broken into two equal sections of* 4 *units).***

I have two different scale factors this time. The horizontal distance will be enlarged, and the vertical distance will be reduced.

2. Create a scale drawing of the original drawing given below using a horizontal scale factor of 150% and a vertical scale factor of 50%. Write numerical equations to find the horizontal and vertical distances.

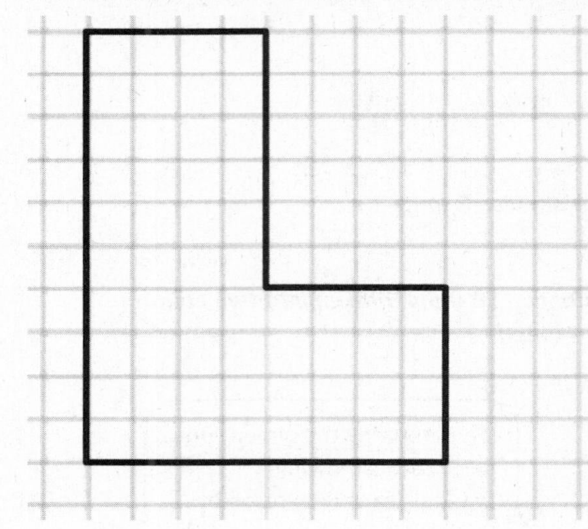

Horizontal Scale Factor: $150\% = \frac{150}{100} = \frac{3}{2}$

Vertical Scale Factor: $50\% = \frac{50}{100} = \frac{1}{2}$

Horizontal Distance: 8 ***units***

The top is broken into two sections of 4 ***units.***

Vertical Distance: 10 ***units***

The right side is broken into two sections, where one section is 6 ***units and the other is*** 4 ***units.***

Horizontal Distance of Scale Drawing:

$(8 \text{ units})(150\%) = (8 \text{ units})\left(\frac{3}{2}\right) = 12 \text{ units}$

However, the top will be broken into two sections of 6 ***units.***

When finding the distances for the scale drawing, I need to make sure I use the correct scale factor for the horizontal distances and the vertical distances.

Vertical Distance of Scale Drawing:

$(10 \text{ units})(50\%) = (10 \text{ units})\left(\frac{1}{2}\right) = 5 \text{ units}$

However, the right side will be broken into two sections, where one section is $(6 \text{ units})\left(\frac{1}{2}\right) = 3 \text{ units}$ ***and the other section is*** $(4 \text{ units})\left(\frac{1}{2}\right) = 2 \text{ units}$.

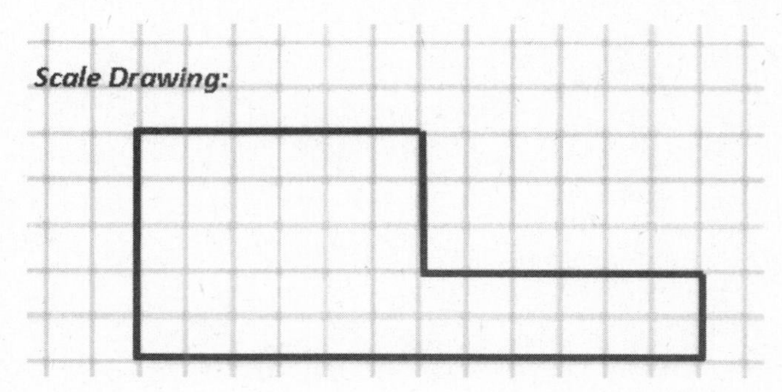

1. Use the diagram below to create a scale drawing using a scale factor of $133\frac{1}{3}\%$. Write numerical equations to find the horizontal and vertical distances in the scale drawing.

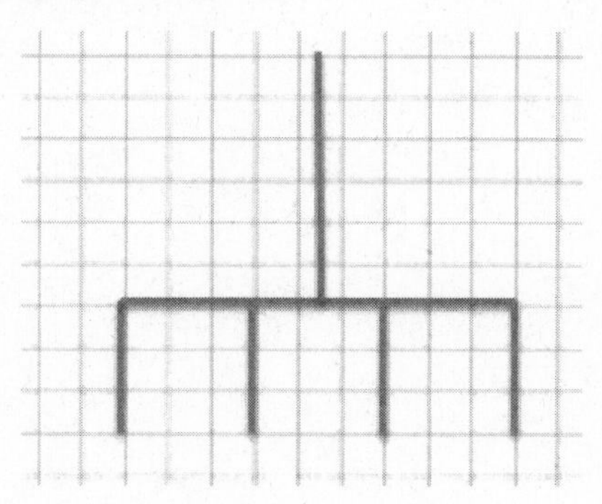

2. Create a scale drawing of the original drawing given below using a horizontal scale factor of 80% and a vertical scale factor of 175%. Write numerical equations to find the horizontal and vertical distances.

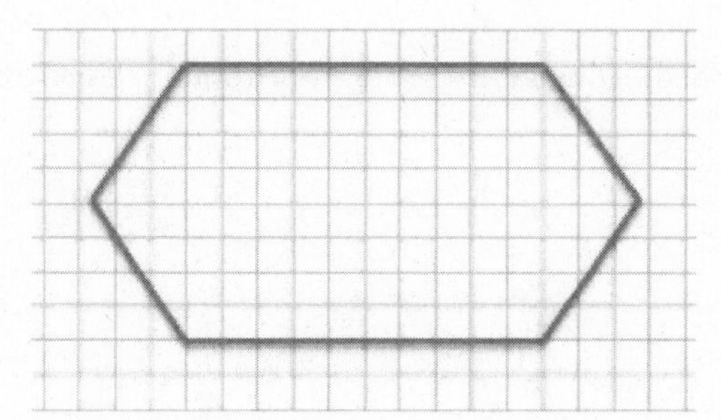

3. The accompanying diagram shows that the length of a pencil from its eraser to its tip is 7 units and that the eraser is 1.5 units wide. The picture was placed on a photocopy machine and reduced to $66\frac{2}{3}\%$. Find the new size of the pencil, and sketch a drawing. Write numerical equations to find the new dimensions.

1.5 units

7 units

4. Use the diagram to answer each question.

   a. What are the corresponding horizontal and vertical distances in a scale drawing if the scale factor is 25%? Use numerical equations to find your answers.

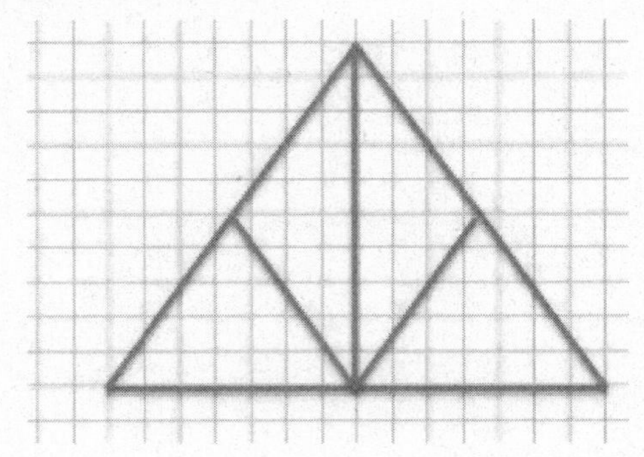

   b. What are the corresponding horizontal and vertical distances in a scale drawing if the scale factor is 160%? Use a numerical equation to find your answers.

5. Create a scale drawing of the original drawing below using a horizontal scale factor of 200% and a vertical scale factor of 250%.

6. Using the diagram below, on grid paper sketch the same drawing using a horizontal scale factor of 50% and a vertical scale factor of 150%.

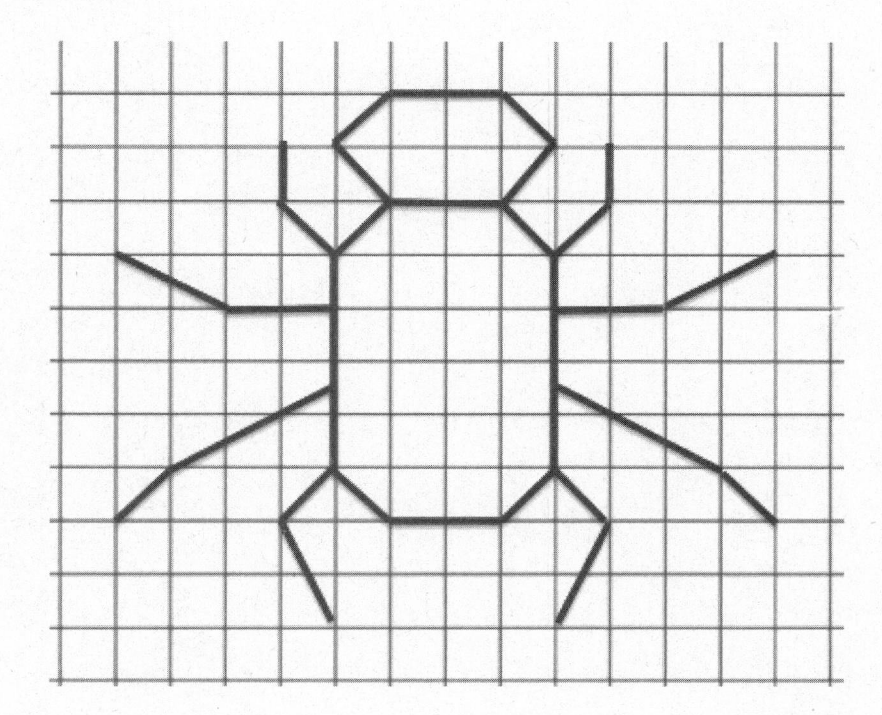

Opening Exercise

Scale factor: $\dfrac{\text{length in SCALE drawing}}{\text{Corresponding length in ORIGINAL drawing}}$

Describe, using percentages, the difference between a reduction and an enlargement.

Use the two drawings below to complete the chart. Calculate the first row (Drawing 1 to Drawing 2) only.

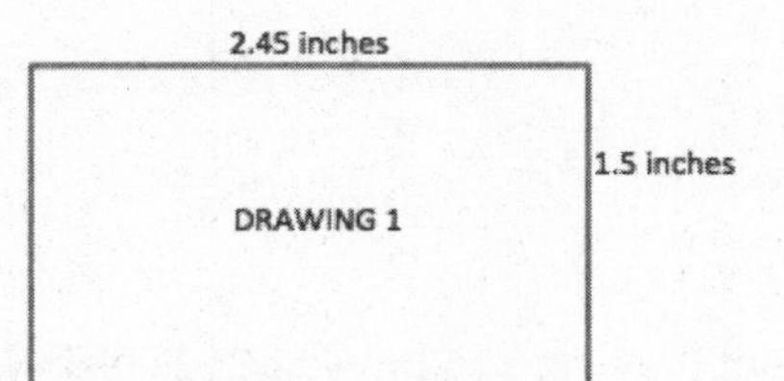

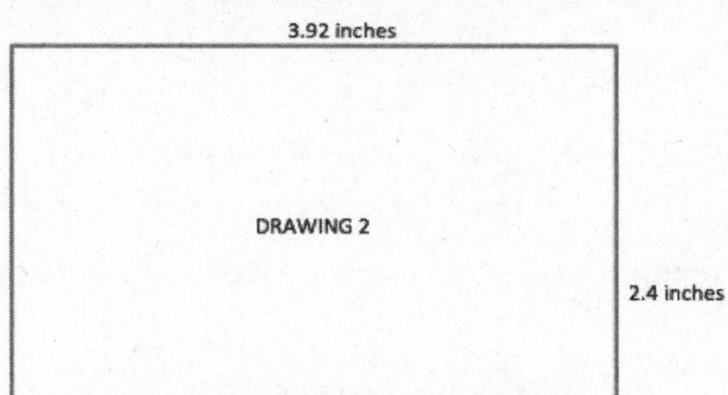

	Quotient of Corresponding Horizontal Distances	Quotient of Corresponding Vertical Distances	Scale Factor as a Percent	Reduction or Enlargement?
Drawing 1 to Drawing 2				
Drawing 2 to Drawing 1				

Compare Drawing 2 to Drawing 1. Using the completed work in the first row, make a conjecture (statement) about what the second row of the chart will be. Justify your conjecture without computing the second row.

Compute the second row of the chart. Was your conjecture proven true? Explain how you know.

Example 1

The scale factor from Drawing 1 to Drawing 2 is 60%. Find the scale factor from Drawing 2 to Drawing 1. Explain your reasoning.

Example 2

A regular octagon is an eight-sided polygon with side lengths that are all equal. All three octagons are scale drawings of each other. Use the chart and the side lengths to compute each scale factor as a percent. How can we check our answers?

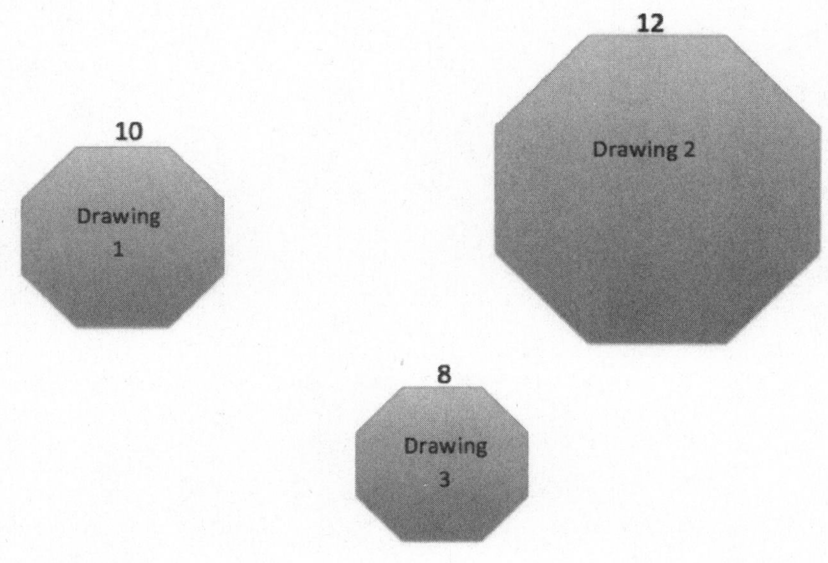

Actual Drawing to Scale Drawing	Scale Factor	Equation to Illustrate Relationship
Drawing 1 to Drawing 2		
Drawing 1 to Drawing 3		
Drawing 2 to Drawing 1		

Drawing 2 to Drawing 3		
Drawing 3 to Drawing 1		
Drawing 3 to Drawing 2		

Example 3

The scale factor from Drawing 1 to Drawing 2 is 112%, and the scale factor from Drawing 1 to Drawing 3 is 84%. Drawing 2 is also a scale drawing of Drawing 3. Is Drawing 2 a reduction or an enlargement of Drawing 3? Justify your answer using the scale factor. The drawing is not necessarily drawn to scale.

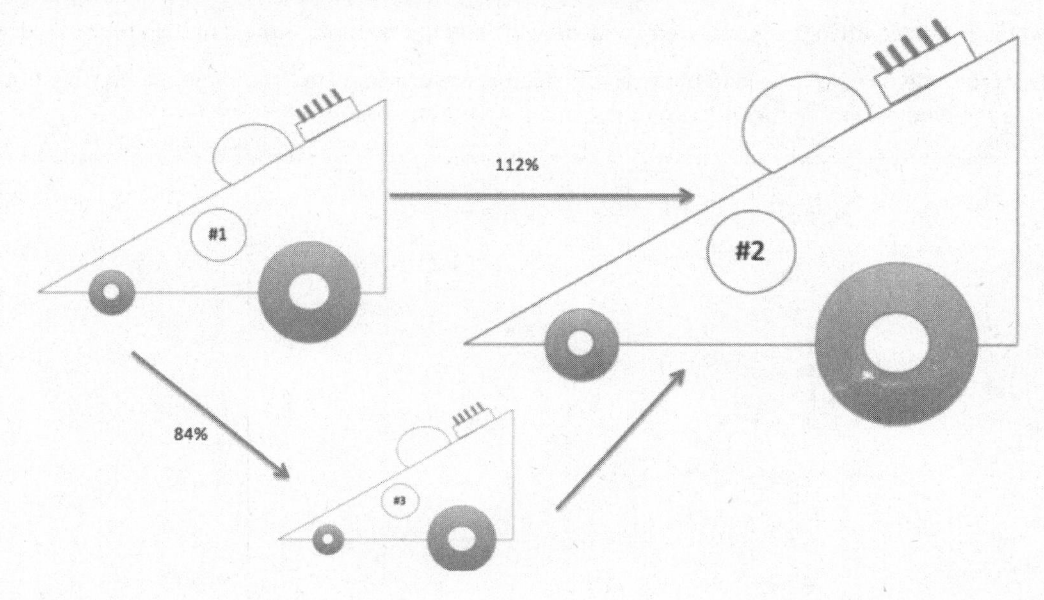

Explain how you could use the scale factors from Drawing 1 to Drawing 2 (112%) and from Drawing 2 to Drawing 3 (75%) to show that the scale factor from Drawing 1 to Drawing 3 is 84%.

Lesson Summary

To compute the scale factor from one drawing to another, use the representation:

$$\text{Quantity} = \text{Percent} \times \text{Whole}$$

where the whole is the length in the actual or original drawing and the quantity is the length in the scale drawing.

If the lengths of the sides are not provided but two scale factors are provided, use the same relationship but use the scale factors as the whole and quantity instead of the given measurements.

Name ______________________________ Date ______________

1. Compute the scale factor, as a percent, for each given relationship. When necessary, round your answer to the nearest tenth of a percent.

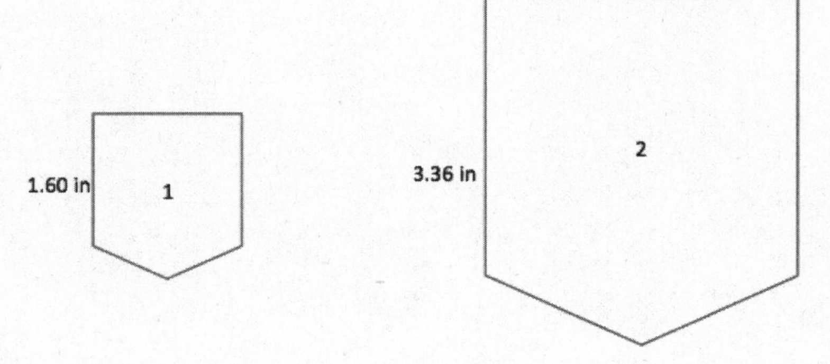

a. Drawing 1 to Drawing 2

b. Drawing 2 to Drawing 1

c. Write two different equations that illustrate how each scale factor relates to the lengths in the diagram.

2. Drawings 2 and 3 are scale drawings of Drawing 1. The scale factor from Drawing 1 to Drawing 2 is 75%, and the scale factor from Drawing 2 to Drawing 3 is 50%. Find the scale factor from Drawing 1 to Drawing 3.

1. The scale factor from Drawing 1 to Drawing 2 is 150%. Justify why Drawing 1 is a scale drawing of Drawing 2 and why it is a reduction of Drawing 2. Include the scale factor in your justification.

Drawing 1

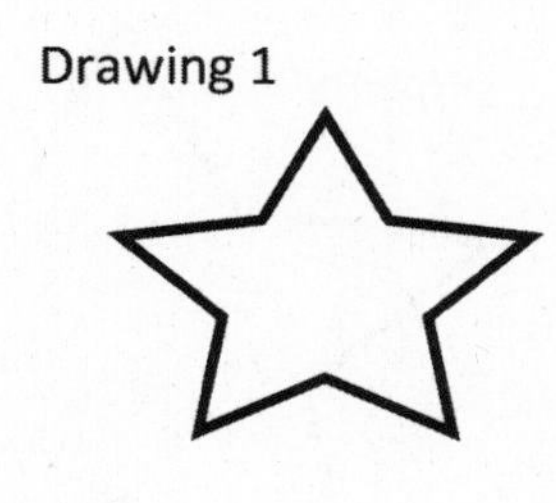

Drawing 2

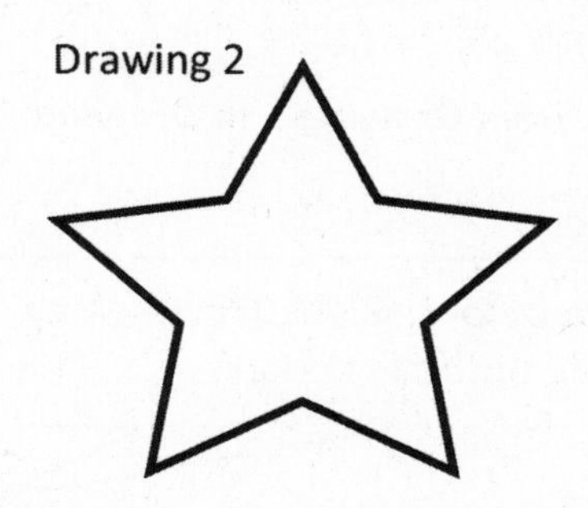

$$\textbf{Length of Drawing 1} = \textbf{Percent} \times \textbf{Length of Drawing 2}$$
$$100\% = \textbf{Percent} \times 150\%$$
$$\left(\frac{1}{150\%}\right)(100\%) = \left(\frac{1}{150\%}\right)(\textbf{Percent} \times 150\%)$$
$$66\frac{2}{3} = \textbf{Percent}$$

The lengths of the original drawings always represent 100%.

Drawing 1 is a scale drawing of Drawing 2 because the lengths of Drawing 1 would be smaller than the corresponding lengths in Drawing 2.

Since the scale factor from Drawing 2 to Drawing 1 is $66\frac{2}{3}\%$, Drawing 1 is a reduction of Drawing 2.

I know a scale drawing is a reduction when the scale factor is less than 100%. If the scale factor is greater than 100%, then the scale drawing is an enlargement.

2. The scale factor from Drawing 2 (presented in the first problem) to Drawing 3 is 125%. What is the scale factor from Drawing 1 to Drawing 3? Explain your reasoning, and check your answer using an example.

$(150\%)(125\%) = (1.50)(1.25) = 1.875$

Therefore, the scale factor from Drawing 1 to Drawing 3 is 187.5%.

Drawing 3

I can choose any length for the length in Drawing 1 when I am checking the scale factors.

Check: Assume that one of the lengths in Drawing 1 is 8 cm.

$(8 \text{ cm})(1.50) = 12 \text{ cm}$

The corresponding length in Drawing 2 would be 12 cm.

$(12 \text{ cm})(1.25) = 15 \text{ cm}$

The corresponding length in Drawing 3 would be 15 cm.

I use the scale factor for Drawing 1 to Drawing 2 and then take the new length and use the scale factor for Drawing 2 to Drawing 3. The result will be the corresponding side length for Drawing 3.

$(8 \text{ cm})\ (1.875) = 15 \text{ cm}$

If I use the scale factor from Drawing 1 to Drawing 3, I find that the corresponding length in Drawing 3 will still be 15 cm.

I use the same length that I originally chose for Drawing 1 and use the scale factor for Drawing 1 to Drawing 3 to determine if the side length for Drawing 3 is the same as when I used the two different scale factors.

3. Cooper drew a picture in the size of a 6-inch by 6-inch square. He wanted to enlarge the original drawing to a size of 7 inches by 7 inches and 11 inches by 11 inches.

   a. Sketch the different sizes of the drawing.

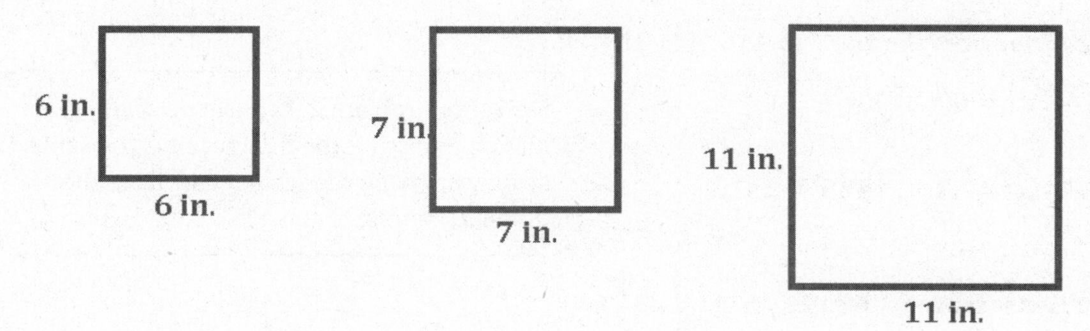

   b. What was the scale factor from the original drawing to the drawing that is 7 inches by 7 inches?

   $\frac{7}{6} = 1.1\overline{6} = 116\frac{2}{3}\%$

   The scale factor from the original drawing to the drawing that is 7 inches by 7 inches was $116\frac{2}{3}\%$.

   I can check to make sure my answer makes sense because I know the 7×7 drawing is an enlargement of the original drawing. Therefore, the scale factor must be greater than 100%.

   c. What was the scale factor from the original drawing to the drawing that is 11 inches by 11 inches?

   $\frac{11}{6} = 1.8\overline{3} = 183\frac{1}{3}\%$

   The scale factor from the original drawing to the drawing that is 11 inches by 11 inches was $183\frac{1}{3}\%$.

   d. What was the scale factor from the 7×7 drawing to the 11×11 drawing?

   $\frac{11}{7} = 1.\overline{571428} = 157\frac{1}{7}\%$

   The scale factor from the 7×7 drawing to the 11×11 drawing is $157\frac{1}{7}\%$.

e. Write an equation to verify how the scale factor from the original drawing to the enlarged 11×11 drawing can be calculated using the scale factors from the original drawing to the 7×7 drawing.

Scale factor from original to 7×7: $\mathbf{116\frac{2}{3}\%}$

Scale factor from the 7×7 ***to*** 11×11: $\mathbf{157\frac{1}{7}\%}$

$$6\left(116\frac{2}{3}\%\right) = 6(1.1\overline{6}) = 7$$

$$7\left(157\frac{1}{7}\%\right) = 7(1.\overline{571428}) = 11$$

Similar to Problem 2, I apply two scale factors and see if the final result is the same as only applying the scale factor from the original drawing to the 11×11 drawing.

Scale Factor from 6×6 ***to*** 11×11: $\mathbf{183\frac{1}{3}\%}$

$$6\left(183\frac{1}{3}\%\right) = 6(1.8\overline{3}) = 11$$

This verifies that the scale factor of $183\frac{1}{3}\%$ ***is equivalent to a scale factor of*** $116\frac{2}{3}\%$ ***followed by a scale factor of*** $157\frac{1}{7}\%$.

1. The scale factor from Drawing 1 to Drawing 2 is $41\frac{2}{3}\%$. Justify why Drawing 1 is a scale drawing of Drawing 2 and why it is an enlargement of Drawing 2. Include the scale factor in your justification.

2. The scale factor from Drawing 1 to Drawing 2 is 40%, and the scale factor from Drawing 2 to Drawing 3 is 37.5%. What is the scale factor from Drawing 1 to Drawing 3? Explain your reasoning, and check your answer using an example.

3. Traci took a photograph and printed it to be a size of 4 units by 4 units as indicated in the diagram. She wanted to enlarge the original photograph to a size of 5 units by 5 units and 10 units by 10 units.

   a. Sketch the different sizes of photographs.

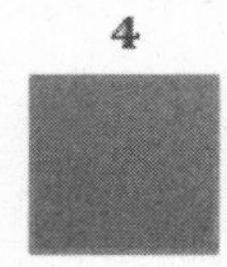

   b. What was the scale factor from the original photo to the photo that is 5 units by 5 units?

   c. What was the scale factor from the original photo to the photo that is 10 units by 10 units?

   d. What was the scale factor from the 5×5 photo to the 10×10 photo?

   e. Write an equation to verify how the scale factor from the original photo to the enlarged 10×10 photo can be calculated using the scale factors from the original to the 5×5, and then from the 5×5 to the 10×10.

4. The scale factor from Drawing 1 to Drawing 2 is 30%, and the scale factor from Drawing 1 to Drawing 3 is 175%. What are the scale factors of each given relationship? Then, answer the question that follows. Drawings are not to scale.

   a. Drawing 2 to Drawing 3

   b. Drawing 3 to Drawing 1

   c. Drawing 3 to Drawing 2

   d. How can you check your answers?

Example 1

The distance around the entire small boat is 28.4 units. The larger figure is a scale drawing of the smaller drawing of the boat. State the scale factor as a percent, and then use the scale factor to find the distance around the scale drawing.

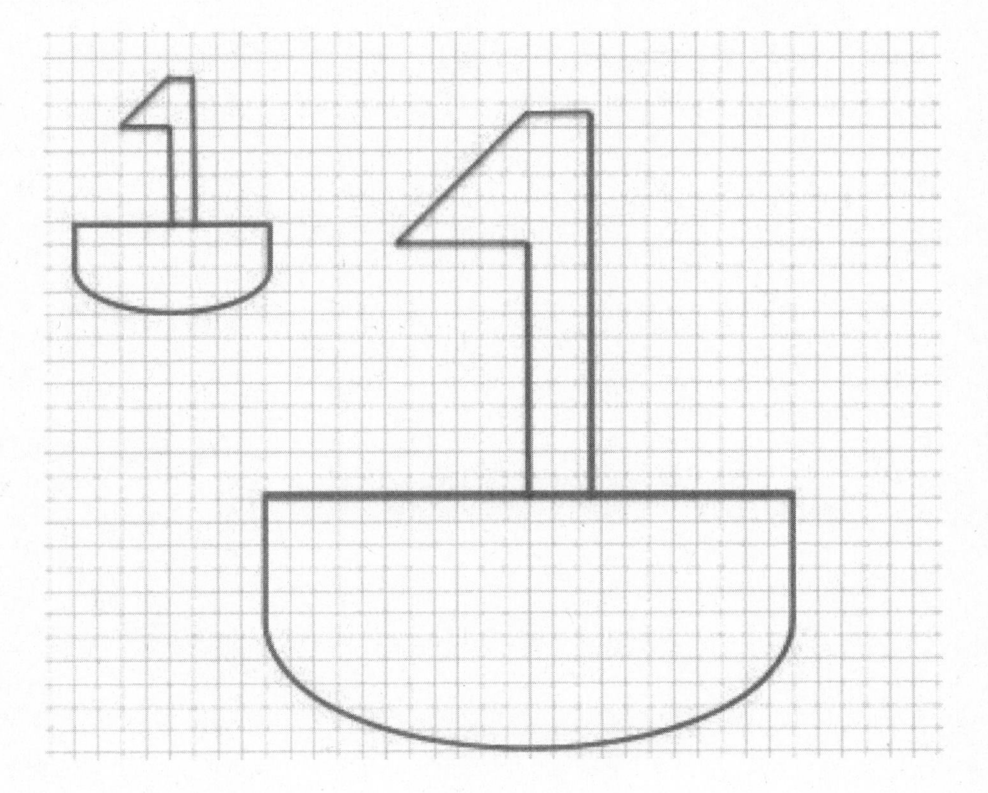

Exercise 1

The length of the longer path is 32.4 units. The shorter path is a scale drawing of the longer path. Find the length of the shorter path, and explain how you arrived at your answer.

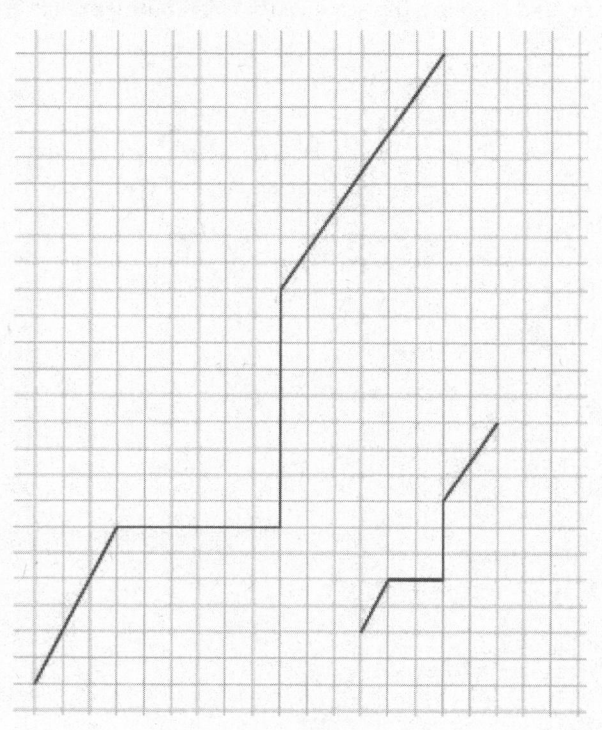

Example 2: Time to Garden

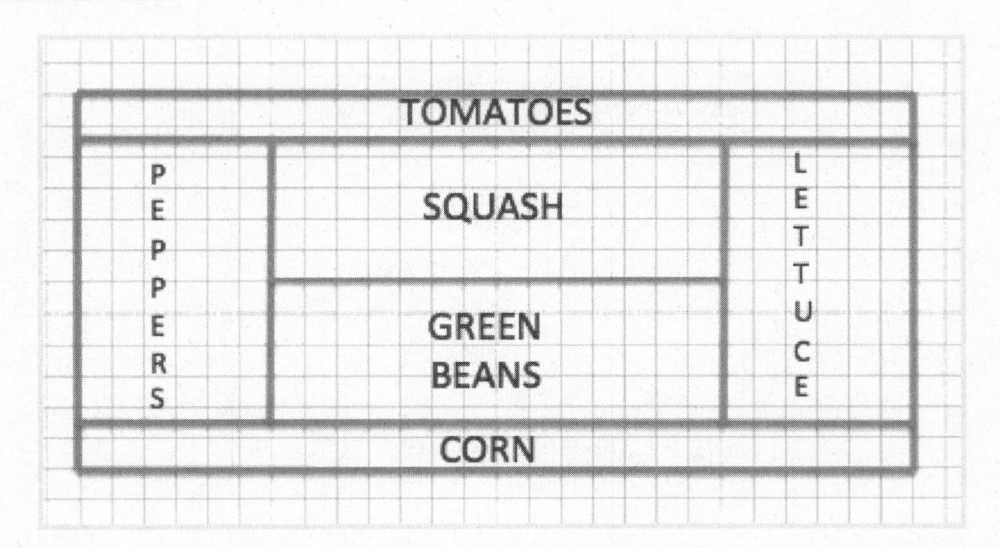

Sherry designed her garden as shown in the diagram above. The distance between any two consecutive vertical grid lines is 1 foot, and the distance between any two consecutive horizontal grid lines is also 1 foot. Therefore, each grid square has an area of one square foot. After designing the garden, Sherry decided to actually build the garden 75% of the size represented in the diagram.

a. What are the outside dimensions shown in the blueprint?

b. What will the overall dimensions be in the actual garden? Write an equation to find the dimensions. How does the problem relate to the scale factor?

c. If Sherry plans to use a wire fence to divide each section of the garden, how much fence does she need?

d. If the fence costs $3.25 per foot plus 7% sales tax, how much would the fence cost in total?

Example 3

Race Car #2 is a scale drawing of Race Car #1. The measurement from the front of Race Car #1 to the back of Race Car #1 is 12 feet, while the measurement from the front of Race Car #2 to the back of Race Car #2 is 39 feet. If the height of Race Car #1 is 4 feet, find the scale factor, and write an equation to find the height of Race Car #2. Explain what each part of the equation represents in the situation.

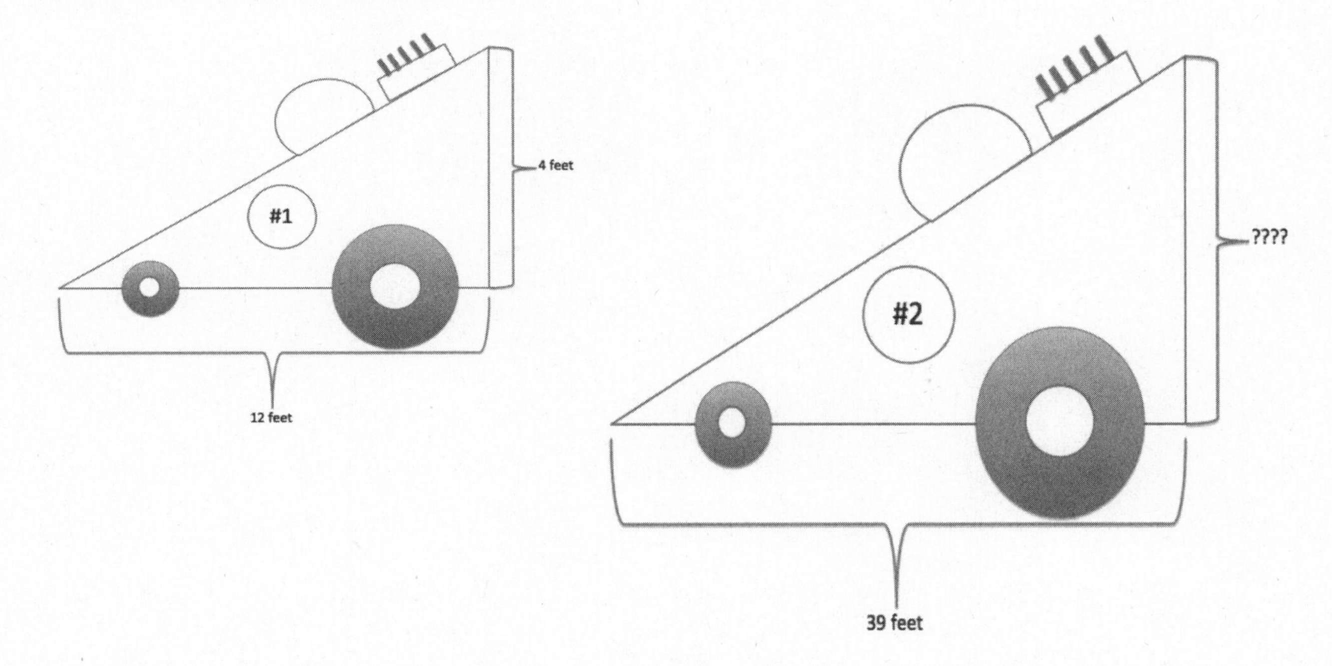

Exercise 2

Determine the scale factor, and write an equation that relates the height of side A in Drawing 1 and the height of side B in Drawing 2 to the scale factor. The height of side A is 1.1 cm. Explain how the equation illustrates the relationship.

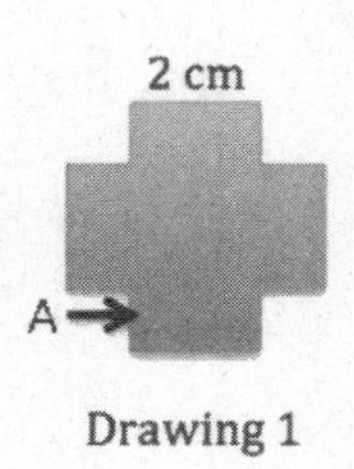

Drawing 1

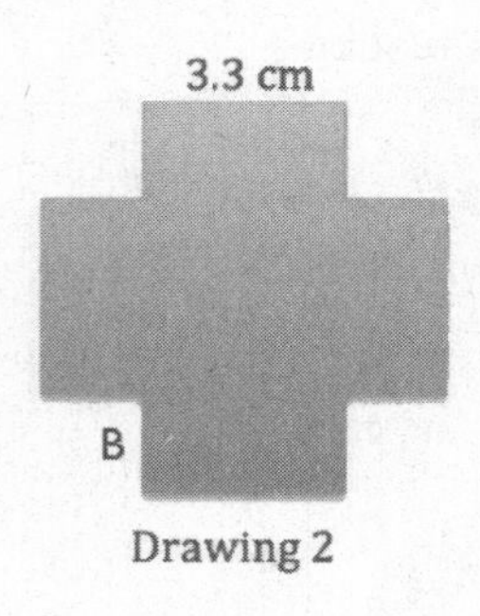

Drawing 2

Exercise 3

The length of a rectangular picture is 8 inches, and the picture is to be reduced to be $45\frac{1}{2}\%$ of the original picture. Write an equation that relates the lengths of each picture. Explain how the equation illustrates the relationship.

Lesson Summary

The scale factor is the number that determines whether the new drawing is an enlargement or a reduction of the original. If the scale factor is greater than 100%, then the resulting drawing is an enlargement of the original drawing. If the scale factor is less than 100%, then the resulting drawing is a reduction of the original drawing.

To compute actual lengths from a scale drawing, a scale factor must first be determined. To do this, use the relationship $\text{Quantity} = \text{Percent} \times \text{Whole}$, where the original drawing represents the whole and the scale drawing represents the quantity. Once a scale factor is determined, then the relationship $\text{Quantity} = \text{Percent} \times \text{Whole}$ can be used again using the scale factor as the percent, the actual length from the original drawing as the whole, and the actual length of the scale drawing as the quantity.

Name ____________________________________ Date ____________________

Each of the designs shown below is to be displayed in a window using strands of white lights. The smaller design requires 225 feet of lights. How many feet of lights does the enlarged design require? Support your answer by showing all work and stating the scale factor used in your solution.

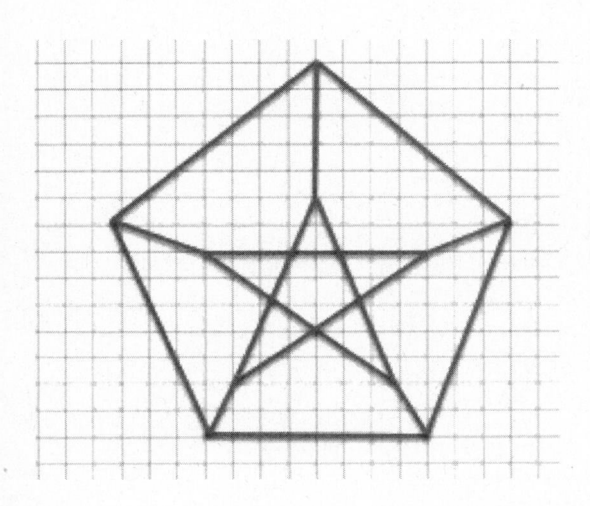

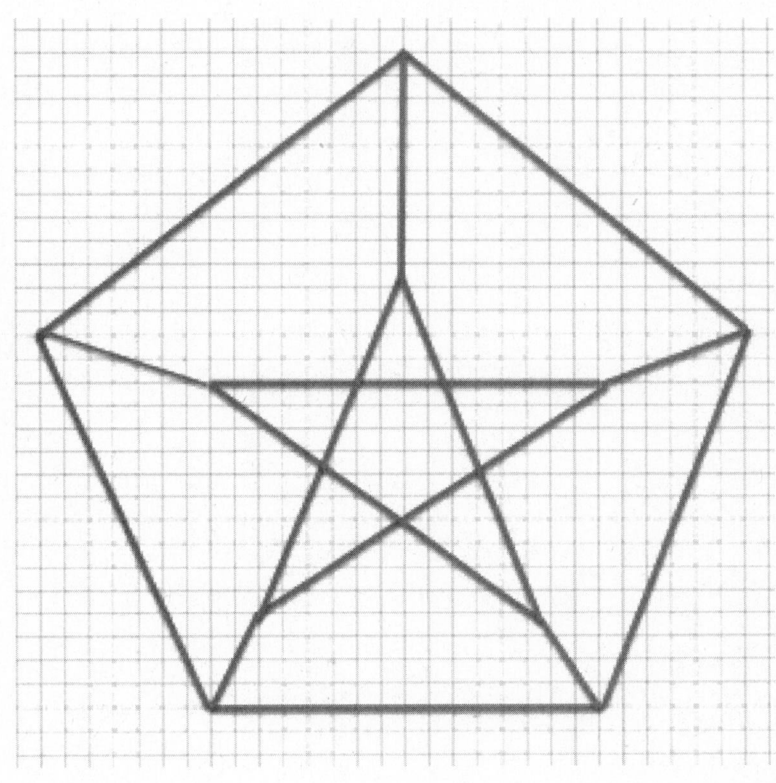

1. A drawing of a toy car is a two-dimensional scale drawing of an actual toy car. The length of the drawing is 1.8 inches, and the width is 0.55 inches. If the length of an actual toy car is 6.12 inches, use an equation to find the width of the actual toy car.

The length of the drawing of the toy car is $\mathbf{1.8}$ ***inches.***

The length of the actual toy car is $\mathbf{6.12}$ ***inches.***

> I can use the corresponding lengths to create an equation that will tell me the scale factor that relates the drawing to the actual toy car.

Scale factor:

$$\mathbf{6.12 = Percent \times 1.8}$$

$$\mathbf{\left(\frac{1}{1.8}\right)(6.12) = Percent \times 1.8\left(\frac{1}{1.8}\right)}$$

$$\mathbf{3.4 = Percent}$$

$$\mathbf{3.4 = 340\%}$$

> The length of the actual toy car is larger than the length of the drawing. Therefore, the scale factor is greater than 1, and the percent is over 100%.

Width of the actual toy car:

$$\mathbf{(0.55)(3.4) = 1.87}$$

> I can use the scale factor to determine the width of the actual toy car.

The width of the actual toy car is $\mathbf{1.87}$ ***inches.***

2. The larger quadrilateral is a scale drawing of the smaller quadrilateral. If the distance around the larger quadrilateral is 36.18 units, what is the distance around the smaller quadrilateral? Use an equation to find the distance, and interpret your solution in the context of the problem.

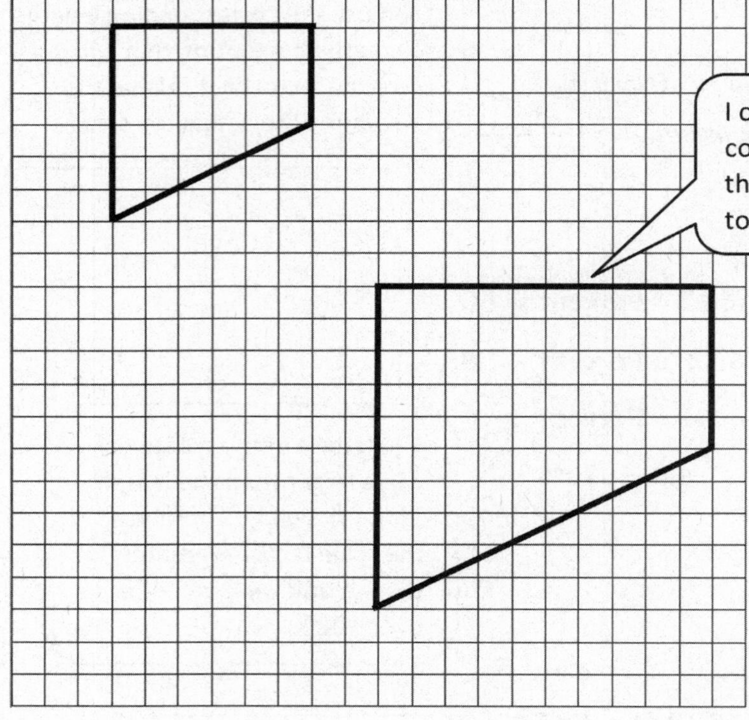

I can use the drawings and count the boxes to determine the horizontal length across the top of each quadrilateral.

The horizontal distance of the smaller quadrilateral is 6 units.

The horizontal distance of the larger quadrilateral is 10 units.

Scale factor:

I can write an equation to determine the scale factor. Once I know the scale factor of the horizontal distances, I am able to determine the distance around the smaller quadrilateral.

$$6 = \text{Percent} \times 10$$

$$\left(\frac{1}{10}\right)(6) = \text{Percent} \times 10\left(\frac{1}{10}\right)$$

$$0.6 = \text{Percent}$$

$$0.6 = 60\%$$

The distance around the smaller quadrilateral:

$$(36.18)(0.6) = 21.708$$

The lengths of the smaller object are 60% of the lengths of the larger object. So I will multiply the distance around the larger object by 60% to determine the distance around the smaller object.

The distance around the smaller quadrilateral is 21.708 units.

1. The smaller train is a scale drawing of the larger train. If the length of the tire rod connecting the three tires of the larger train, as shown below, is 36 inches, write an equation to find the length of the tire rod of the smaller train. Interpret your solution in the context of the problem.

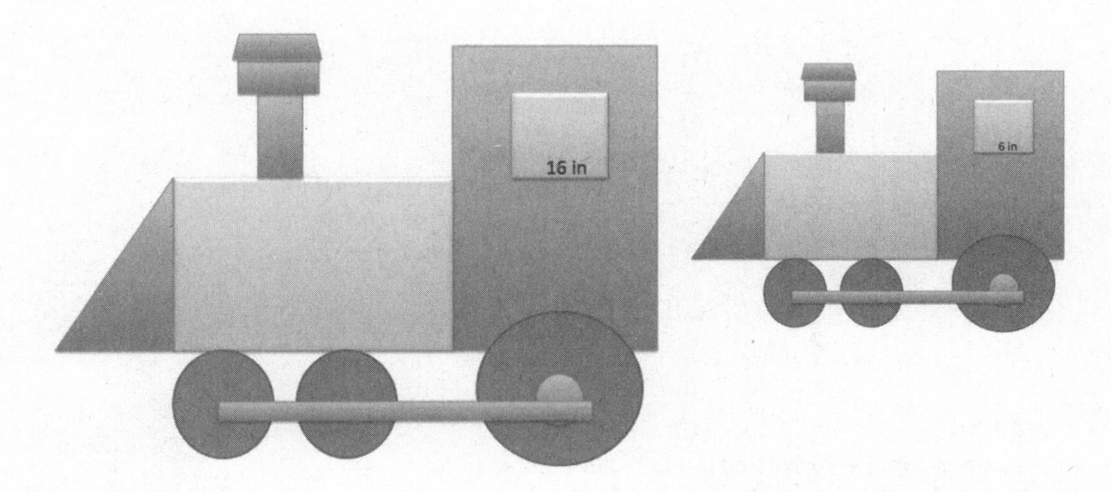

2. The larger arrow is a scale drawing of the smaller arrow. The distance around the smaller arrow is 25.66 units. What is the distance around the larger arrow? Use an equation to find the distance and interpret your solution in the context of the problem.

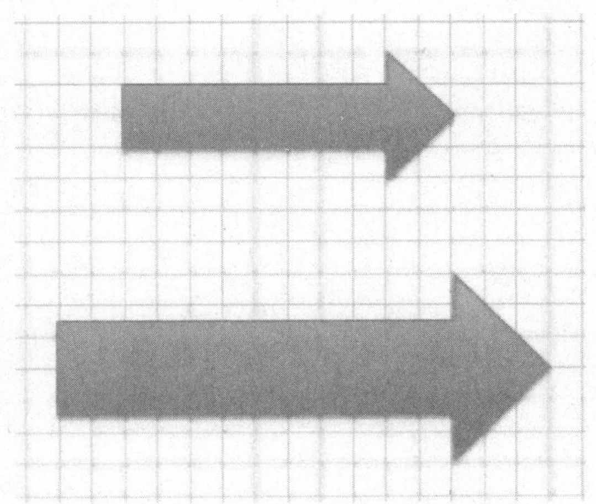

3. The smaller drawing below is a scale drawing of the larger. The distance around the larger drawing is 39.4 units. Using an equation, find the distance around the smaller drawing.

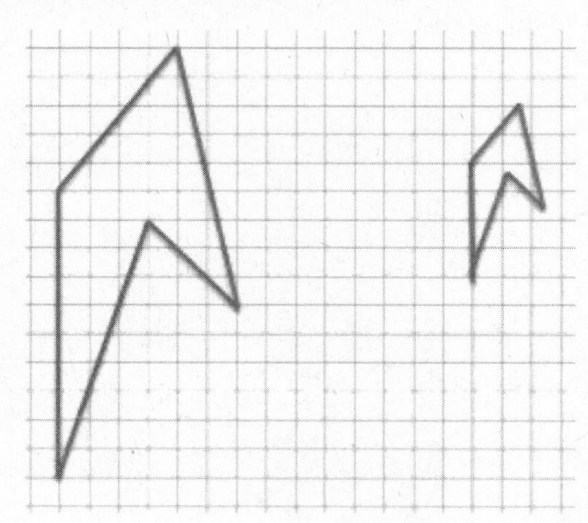

4. The figure is a diagram of a model rocket and is a two-dimensional scale drawing of an actual rocket. The length of a model rocket is 2.5 feet, and the wing span is 1.25 feet. If the length of an actual rocket is 184 feet, use an equation to find the wing span of the actual rocket.

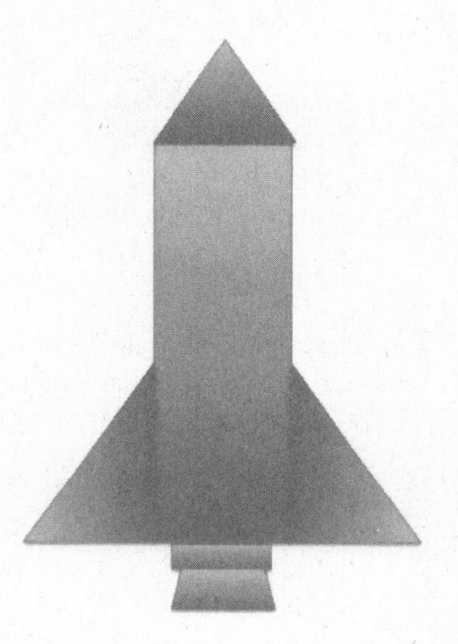

Opening Exercise

For each diagram, Drawing 2 is a scale drawing of Drawing 1. Complete the accompanying charts. For each drawing, identify the side lengths, determine the area, and compute the scale factor. Convert each scale factor into a fraction and percent, examine the results, and write a conclusion relating scale factors to area.

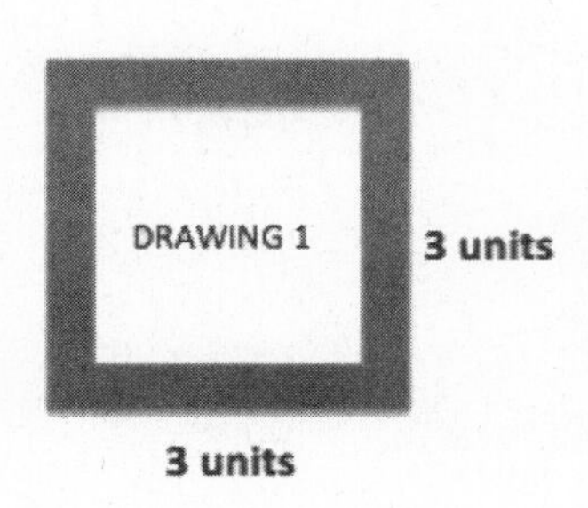

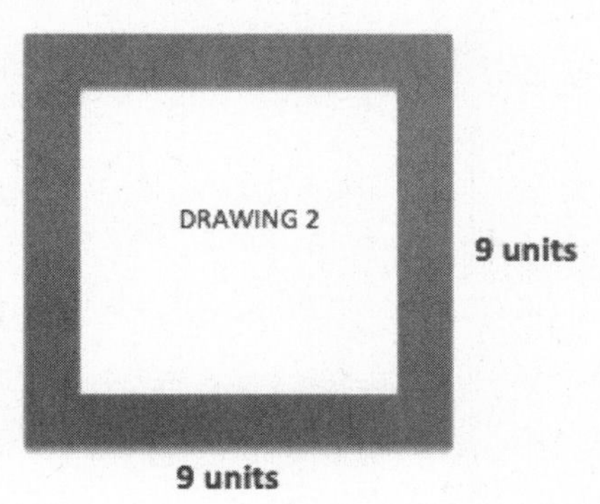

	Drawing 1	Drawing 2	Scale Factor as a Fraction and Percent
Side			
Area (sq. units)			

Scale Factor: ____________________

Quotient of Areas: ______________________

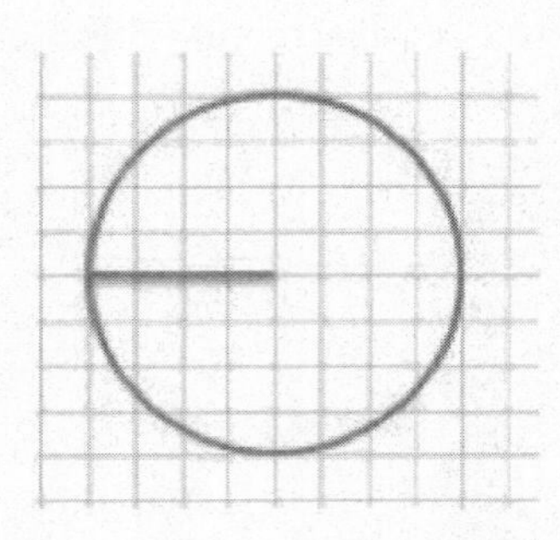

DRAWING 1

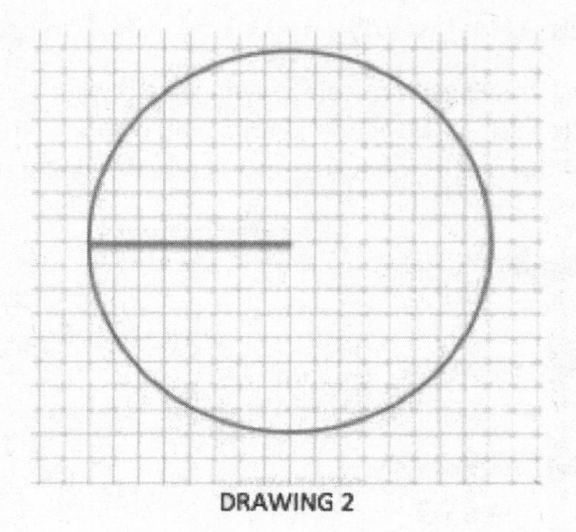

DRAWING 2

	Drawing 1	Drawing 2	Scale Factor as a Fraction and Percent
Radius			
Area (sq. units)			

Scale Factor: ____________________ Quotient of Areas: ____________________

The length of each side in Drawing 1 is 12 units, and the length of each side in Drawing 2 is 6 units.

Drawing 1

Drawing 2

	Drawing 1	Drawing 2	Scale Factor as a Fraction and Percent
Side			
Area (sq. units)			

Scale Factor: ________________

Quotient of Areas: ________________

Conclusion:

Example 1

What percent of the area of the large square is the area of the small square?

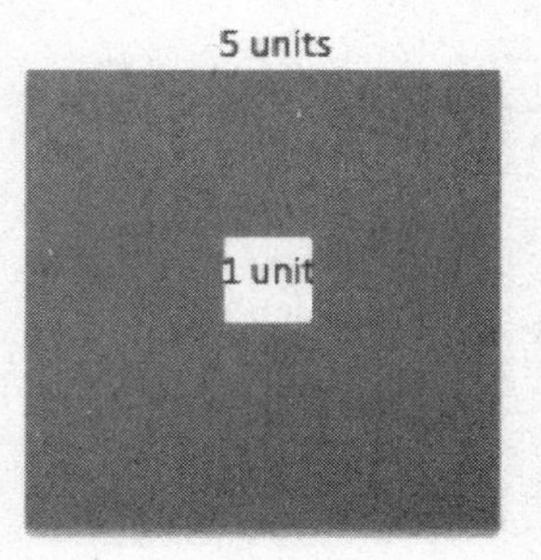

Example 2

What percent of the area of the large disk lies outside the shaded disk?

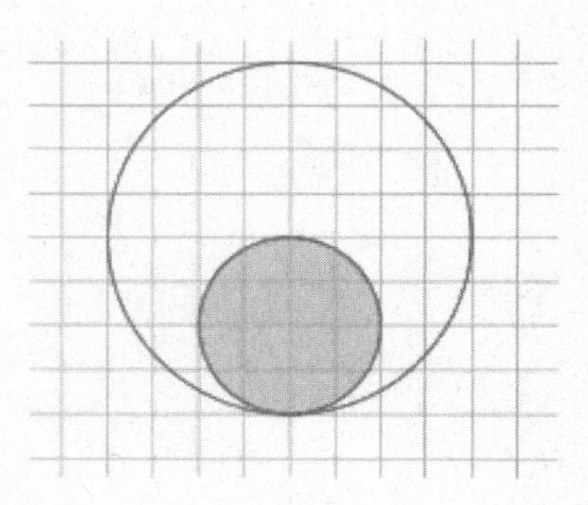

Example 3

If the area of the shaded region in the larger figure is approximately 21.5 square inches, write an equation that relates the areas using scale factor and explain what each quantity represents. Determine the area of the shaded region in the smaller scale drawing.

10 inches

6 inches

Example 4

Use Figure 1 below and the enlarged scale drawing to justify why the area of the scale drawing is k^2 times the area of the original figure.

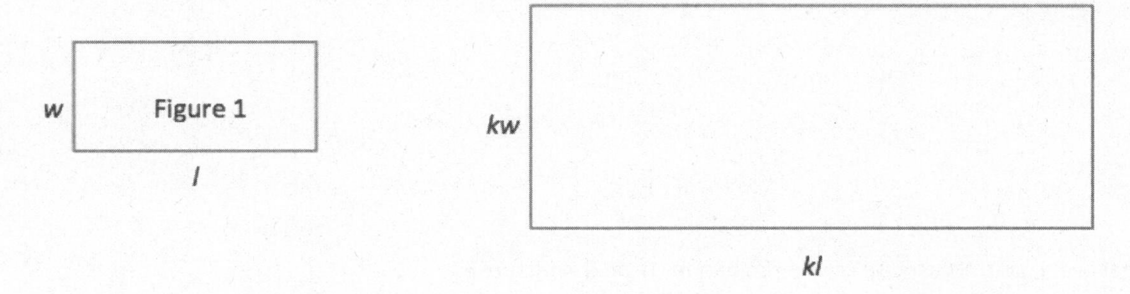

Explain why the expressions $(kl)(kw)$ and k^2lw are equivalent. How do the expressions reveal different information about this situation?

Exercise 1

The Lake Smith basketball team had a team picture taken of the players, the coaches, and the trophies from the season. The picture was 4 inches by 6 inches. The team decided to have the picture enlarged to a poster and then enlarged again to a banner measuring 48 inches by 72 inches.

a. Sketch drawings to illustrate the original picture and enlargements.

b. If the scale factor from the picture to the poster is 500%, determine the dimensions of the poster.

c. What scale factor is used to create the banner from the picture?

d. What percent of the area of the picture is the area of the poster? Justify your answer using the scale factor and by finding the actual areas.

e. Write an equation involving the scale factor that relates the area of the poster to the area of the picture.

f. Assume you started with the banner and wanted to reduce it to the size of the poster. What would the scale factor as a percent be?

g. What scale factor would be used to reduce the poster to the size of the picture?

Lesson Summary

If the scale factor is represented by k, then the area of the scale drawing is k^2 times the corresponding area of the original drawing.

Name ____________________ Date ____________

Write an equation relating the area of the original (larger) drawing to its smaller scale drawing. Explain how you determined the equation. What percent of the area of the larger drawing is the smaller scale drawing?

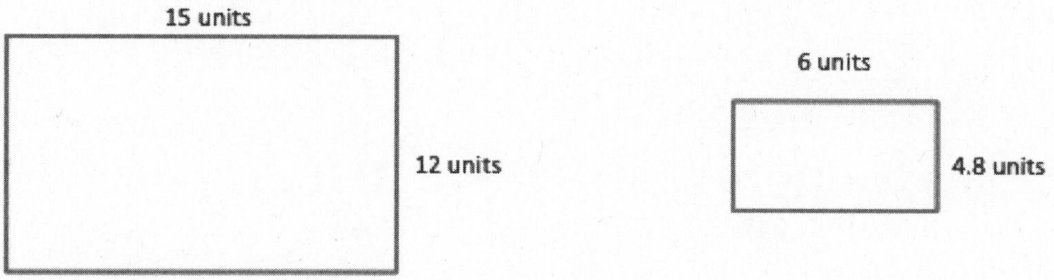

15 units

12 units

6 units

4.8 units

1. Use the diagram of the circles to answer the following questions.

   a. What percent of the area of the larger circle is shaded?

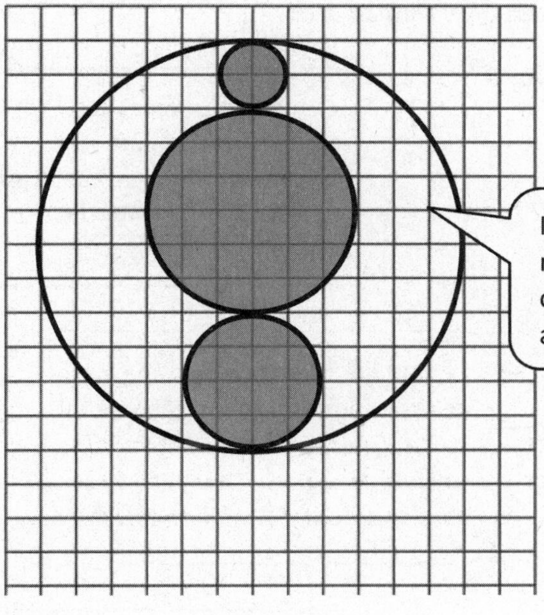

I can use the diagram to determine the radius of each of the circles. There are four circles in this diagram, three on the inside and one outside circle.

Shaded small circle: radius $= 1$ ***unit***

Shaded medium circle: radius $= 2$ ***units***

Shaded large circle: radius $= 3$ ***units***

Outside circle: radius $= 6$ ***units***

Let A ***represent the area of the outside circle.***

Area of small circle: $\left(\frac{1}{6}\right)^2 A = \frac{1}{36}A$

I can compare the area of each shaded circle to the area of the outside circle. Because we are working with area, I need to square the scale factor that compares the side lengths.

Area of medium circle: $\left(\frac{2}{6}\right)^2 A = \frac{4}{36}A$

Area of large circle: $\left(\frac{3}{6}\right)^2 A = \frac{9}{36}A$

Area of shaded region: $\frac{1}{36}A + \frac{4}{36}A + \frac{9}{36}A = \frac{14}{36}A = \frac{14}{36}A \times 100\% = \left(38\frac{8}{9}\%\right)A$

The area of the shaded region is $38\frac{8}{9}\%$ ***of the area of the entire circle.***

b. Using 3.14 as an estimate for π, the area of the outside circle is approximately 113.04 in^2. Determine the area of the shaded region.

If A represents the area of the outside circle, then the total shaded area:

$$\frac{14}{36}A = \frac{14}{36}(113.04) = 43.96$$

The area of the shaded region is approximately 43.96 in^2.

I can use the expression that I came up with in part (a) to determine the area. I just need to replace the A with the actual area of the outside circle.

c. What percent of the outside circle is unshaded?

$$113.04 - 43.96 = 69.08$$

I know the total area and the area of the shaded region. I can use this to determine the area of the unshaded region. Then, I can use that area to determine the percent.

Therefore, the area of the unshaded region is approximately 69.08 in^2.

The percent of the outside circular region that is unshaded is

$$\frac{69.08}{113.04} = 0.6\overline{1} = 61\frac{1}{9}\%.$$

I could have subtracted the percent the shaded region represents ($38\frac{8}{9}\%$) from 100% because I know the entire area represents 100%.

d. What percent of the area of the medium circle is the area of the large circle?

Scale factor: $\frac{3}{2}$

Area: $\left(\frac{3}{2}\right)^2 = \frac{9}{4} = \frac{9}{4} \times 100\% = 225\%$

To determine the percent, I will first write out the scale factor, comparing the radius of the large circle to the radius of the medium circle.

The area of the large circle is 225% of the area of the medium circle.

2. On the square page of a menu shown below, five 3-in. by 3-in. squares are cut out for pictures. If these cut-out regions make up $\frac{5}{36}$ of the area of the entire page, what are the dimensions of the page?

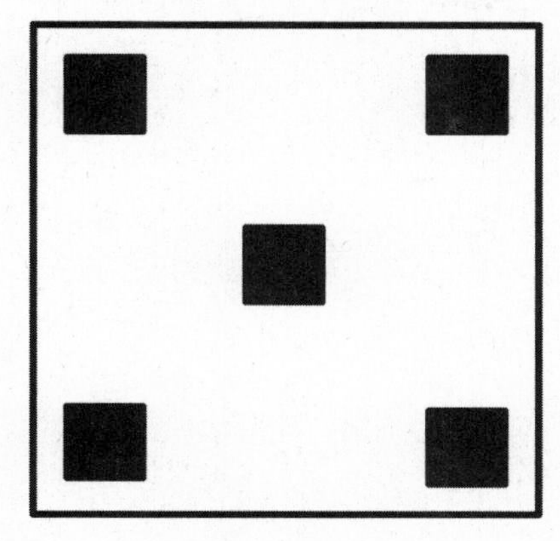

Since the cut-out regions make up $\frac{5}{36}$ ***of the entire page, each cut-out region makes up*** $\frac{\frac{5}{36}}{5} = \frac{1}{36}$ ***of the entire page.***

To figure out what part of the area is covered by just one square, I can divide by the number of squares.

$$\left(\frac{1}{6}\right)^2 = \frac{1}{36}$$

The scale factor is $\frac{1}{6}$.

In this question, I was given information about the areas, but I need to work backward to determine the side lengths. Because the scale factor was squared to determine the areas, I need to determine what number was squared to determine the scale factor and work toward the side lengths.

To find the dimensions of the square page:

$$\textbf{Quantity} = \textbf{Percent} \times \textbf{Whole}$$

$$\textbf{Small square side length} = \textbf{Percent} \times \textbf{Page length}$$

$$3 \text{ in.} = \frac{1}{6} \times \text{Page length}$$

$$6(3 \text{ in.}) = 6\left(\frac{1}{6}\right) \times \text{Page length}$$

$$18 \text{ in.} = \text{Page length}$$

I can write a percent as a fraction to use in the formula.

***The dimensions of the square menu page are* 18 in. *by* 18 in.**

1. What percent of the area of the larger circle is shaded?

   a. Solve this problem using scale factors.

   b. Verify your work in part (a) by finding the actual areas.

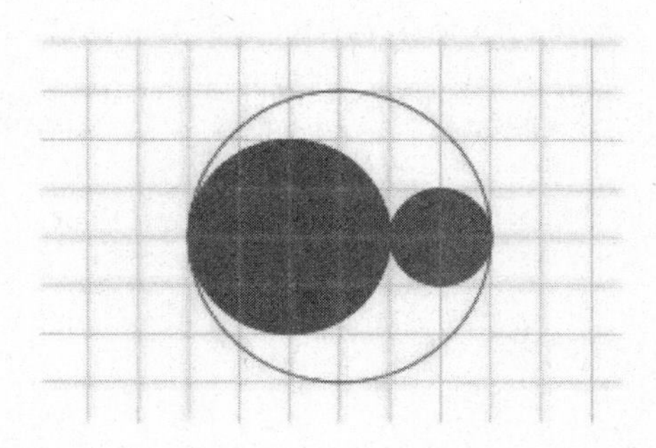

2. The area of the large disk is 50.24 units^2.

   a. Find the area of the shaded region using scale factors. Use 3.14 as an estimate for π.

   b. What percent of the large circular region is unshaded?

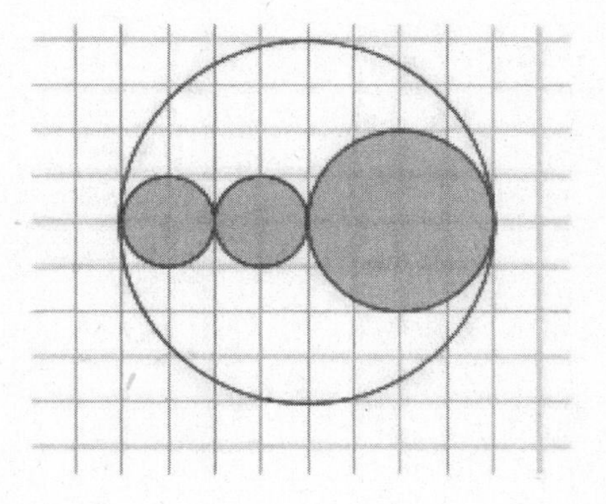

3. Ben cut the following rockets out of cardboard. The height from the base to the tip of the smaller rocket is 20 cm. The height from the base to the tip of the larger rocket is 120 cm. What percent of the area of the smaller rocket is the area of the larger rocket?

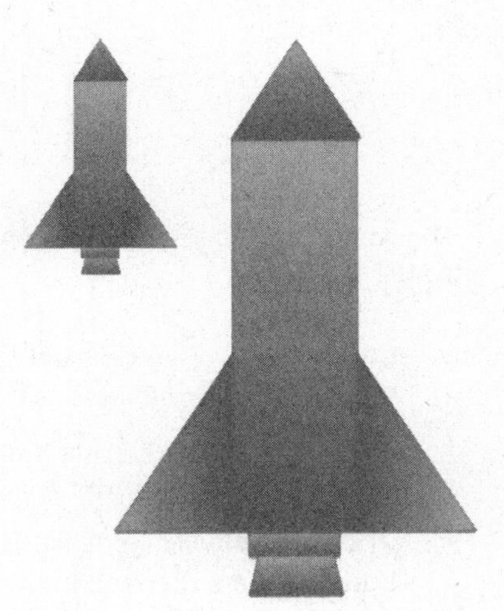

4. In the photo frame depicted below, three 5 inch by 5 inch squares are cut out for photographs. If these cut-out regions make up $\frac{3}{16}$ of the area of the entire photo frame, what are the dimensions of the photo frame?

5. Kelly was online shopping for envelopes for party invitations and saw these images on a website.

The website listed the dimensions of the small envelope as 6 in. by 8 in. and the medium envelope as 10 in. by $13\frac{1}{3}$ in.

a. Compare the dimensions of the small and medium envelopes. If the medium envelope is a scale drawing of the small envelope, what is the scale factor?

b. If the large envelope was created based on the dimensions of the small envelope using a scale factor of 250%, find the dimensions of the large envelope.

c. If the medium envelope was created based on the dimensions of the large envelope, what scale factor was used to create the medium envelope?

d. What percent of the area of the larger envelope is the area of the medium envelope?

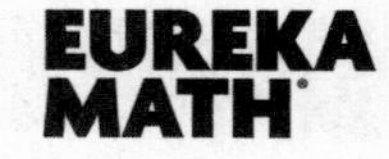

Opening Exercise

Number of girls in classroom:	Number of boys in classroom:	Total number of students in classroom:
Percent of the total number of students that are girls:	Percent of the total number of students that are boys:	Percent of boys and girls in the classroom:
Number of girls whose names start with a vowel:	Number of boys whose names start with a vowel:	Number of students whose names start with a vowel:
Percent of girls whose names start with a vowel:	Percent of boys whose names start with a vowel:	
Percent of the total number of students who are girls whose names start with a vowel:	Percent of the total number of students who are boys whose names start with a vowel:	Percent of students whose names start with a vowel:

Example 1

A school has 60% girls and 40% boys. If 20% of the girls wear glasses and 40% of the boys wear glasses, what percent of all students wears glasses?

Exercise 1

How does the percent of students who wear glasses change if the percent of girls and boys remains the same (that is, 60% girls and 40% boys), but 20% of the boys wear glasses and 40% of the girls wear glasses?

Exercise 2

How would the percent of students who wear glasses change if the percent of girls is 40% of the school and the percent of boys is 60% of the school, and 40% of the girls wear glasses and 20% of the boys wear glasses? Why?

Example 2

The weight of the first three containers is 12% more than the second, and the third container is 20% lighter than the second. By what percent is the first container heavier than the third container?

Exercise 3

Matthew's pet dog is 7% heavier than Harrison's pet dog, and Janice's pet dog is 20% lighter than Harrison's. By what percent is Matthew's dog heavier than Janice's?

Example 3

In one year's time, 20% of Ms. McElroy's investments increased by 5%, 30% of her investments decreased by 5%, and 50% of her investments increased by 3%. By what percent did the total of her investments increase?

Exercise 4

A concert had $6{,}000$ audience members in attendance on the first night and the same on the second night. On the first night, the concert exceeded expected attendance by 20%, while the second night was below the expected attendance by 20%. What was the difference in percent of concert attendees and expected attendees for both nights combined?

Lesson Summary

When solving a percent population problem, you must first define the variable. This gives a reference of what the whole is. Then, multiply the sub-populations (such as girls and boys) by the given category (total students wearing glasses) to find the percent in the whole population.

Name ______________________________ Date ______________

1. Jodie spent 25% less buying her English reading book than Claudia. Gianna spent 9% less than Claudia. Gianna spent more than Jodie by what percent?

2. Mr. Ellis is a teacher who tutors students after school. Of the students he tutors, 30% need help in computer science and the rest need assistance in math. Of the students who need help in computer science, 40% are enrolled in Mr. Ellis's class during the school day. Of the students who need help in math, 25% are enrolled in his class during the school day. What percent of the after-school students are enrolled in Mr. Ellis's classes?

1. During a school fundraiser, 70% of customers ordered vanilla ice cream, and 30% of customers ordered chocolate ice cream. Of the customers who ordered vanilla, 80% asked for sprinkles as well. Of the customers who ordered chocolate, 50% asked for sprinkles. What is the percent of customers who ordered sprinkles with their ice cream?

Let c represent the number of customer orders.

> I don't know how many customers bought ice cream, so I need to use a variable to represent this value.

Vanilla Orders: $70\% \times c = 0.7c$

Chocolate Orders: $30\% \times c = 0.3c$

> I need to multiply the number of orders by the percent of orders that included sprinkles.

Vanilla Orders with Sprinkles: $0.7c \times 0.8 = 0.56c$

Chocolate Orders with Sprinkles: $0.3c \times 0.5 = 0.15c$

> I know the percent of vanilla orders with sprinkles and the percent of chocolate orders with sprinkles; now I need to determine the total percent of orders with sprinkles.

Orders with Sprinkles: $0.56c + 0.15c = 0.71c$

Therefore, the percent of customers who ordered sprinkles is 71%.

2. The city zoo keeps records on the number of guests they have throughout the year. Last year, 40% of the guests were adults, and 60% of the guests were children. This year, there was an 8% decrease in adult guests and a 3% increase in children guests. What is the percent increase or decrease in guests this year?

Let g represent the number of guests last year.

Adults Decrease: $(0.4g)(0.92) = 0.368g$

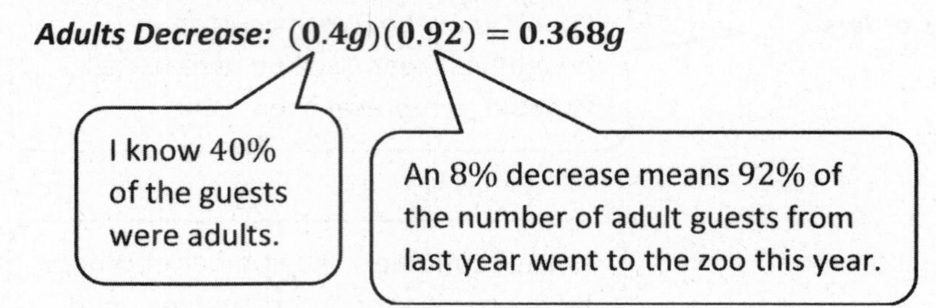

Children Increase: $(0.6g)(1.03) = 0.618g$

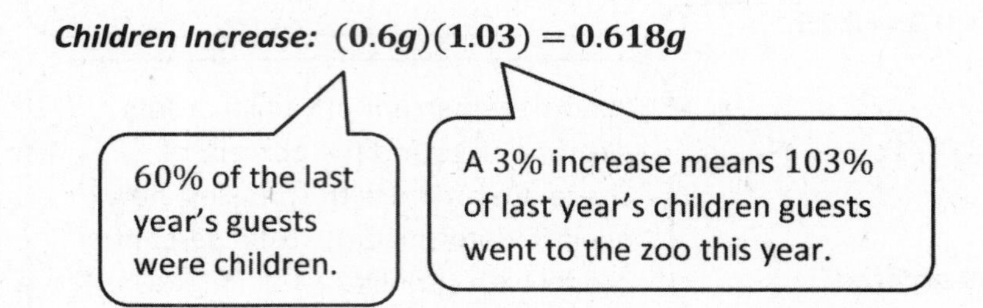

$0.368g + 0.618\,g = 0.986g$

$100\% - 98.6\% = 1.4\%$

Therefore, the number of guests decreased by 1.4%.

3. A seventh grade math class is made up of 40% girls and 60% boys, and 35% of the entire class is earning an A in the class. If 50% of the girls are earning an A, what is the percent of boys who are earning an A?

Let b represent the percent of boys who are earning an A.

$$0.4(0.5) + 0.6b = 1(0.35)$$

I know the total percentage of A's is 35%. Therefore, the sum of percentages of the girls A's and the boys A's is 35%.

$$0.2 + 0.6b = 0.35$$

$$0.2 + 0.6b - 0.2 = 0.35 - 0.2$$

I use the additive inverse of 0.2 in order to isolate the variable.

$$0.6b = 0.15$$

$$\left(\frac{1}{0.6}\right)(0.6b) = \left(\frac{1}{0.6}\right)0.15$$

$$b = 0.25$$

I use the multiplicative inverse of 0.6 in order to isolate the variable.

Therefore, 25% ***of the boys are earning an A.***

1. One container is filled with a mixture that is 30% acid. A second container is filled with a mixture that is 50% acid. The second container is 50% larger than the first, and the two containers are emptied into a third container. What percent of acid is in the third container?

2. The store's markup on a wholesale item is 40%. The store is currently having a sale, and the item sells for 25% off the retail price. What is the percent of profit made by the store?

3. During lunch hour at a local restaurant, 90% of customers order a meat entrée and 10% order a vegetarian entrée. Of the customers who order a meat entrée, 80% order a drink. Of the customers who order a vegetarian entrée, 40% order a drink. What is the percent of customers who order a drink with their entrée?

4. Last year's spell-a-thon spelling test for a first grade class had 15% more words with four or more letters than this year's spelling test. Next year, there will be 5% less than this year. What percent more words have four or more letters in last year's test than next year's?

5. An ice cream shop sells 75% less ice cream in December than in June. Twenty percent more ice cream is sold in July than in June. By what percent did ice cream sales increase from December to July?

6. The livestock on a small farm the prior year consisted of 40% goats, 10% cows, and 50% chickens. This year, there is a 5% decrease in goats, 9% increase in cows, and 15% increase in chickens. What is the percent increase or decrease of livestock this year?

7. In a pet shelter that is occupied by 55% dogs and 45% cats, 60% of the animals are brought in by concerned people who found these animals in the streets. If 90% of the dogs are brought in by concerned people, what is the percent of cats that are brought in by concerned people?

8. An artist wants to make a particular teal color paint by mixing a 75% blue hue and 25% yellow hue. He mixes a blue hue that has 85% pure blue pigment and a yellow hue that has 60% of pure yellow pigment. What is the percent of pure pigment that is in the resulting teal color paint?

9. On Mina's block, 65% of her neighbors do not have any pets, and 35% of her neighbors own at least one pet. If 25% of the neighbors have children but no pets, and 60% of the neighbors who have pets also have children, what percent of the neighbors have children?

Opening Exercise

Imagine you have two equally-sized containers. One is pure water, and the other is 50% water and 50% juice. If you combined them, what percent of juice would be the result?

	1st Liquid	2nd Liquid	Resulting Liquid
Amount of Liquid (gallons)			
Amount of Pure Juice (gallons)			

If a 2-gallon container of pure juice is added to 3 gallons of water, what percent of the mixture is pure juice?

	1st Liquid	2nd Liquid	Resulting Liquid
Amount of Liquid (gallons)			
Amount of Pure Juice (gallons)			

If a 2-gallon container of juice mixture that is 40% pure juice is added to 3 gallons of water, what percent of the mixture is pure juice?

	1st Liquid	2nd Liquid	Resulting Liquid
Amount of Liquid (gallons)			
Amount of Pure Juice (gallons)			

If a 2-gallon juice cocktail that is 40% pure juice is added to 3 gallons of pure juice, what percent of the resulting mixture is pure juice?

	1st Liquid	2nd Liquid	Resulting Liquid
Amount of Liquid (gallons)			
Amount of Pure Juice (gallons)			

Example 1

A 5-gallon container of trail mix is 20% nuts. Another trail mix is added to it, resulting in a 12-gallon container of trail mix that is 40% nuts.

a. Write an equation to describe the relationships in this situation.

b. Explain in words how each part of the equation relates to the situation.

c. What percent of the second trail mix is nuts?

Exercise 1

Represent each situation using an equation, and show all steps in the solution process.

a. A 6-pint mixture that is 25% oil is added to a 3-pint mixture that is 40% oil. What percent of the resulting mixture is oil?

b. An 11-ounce gold chain of 24% gold was made from a melted down 4-ounce charm of 50% gold and a golden locket. What percent of the locket was pure gold?

c. In a science lab, two containers are filled with mixtures. The first container is filled with a mixture that is 30% acid. The second container is filled with a mixture that is 50% acid, and the second container is 50% larger than the first. The first and second containers are then emptied into a third container. What percent of acid is in the third container?

Example 2

Soil that contains 30% clay is added to soil that contains 70% clay to create 10 gallons of soil containing 50% clay. How much of each of the soils was combined?

Exercise 2

The equation $(0.2)(x) + (0.8)(6 - x) = (0.4)(6)$ is used to model a mixture problem.

a. How many units are in the total mixture?

b. What percents relate to the two solutions that are combined to make the final mixture?

c. The two solutions combine to make 6 units of what percent solution?

d. When the amount of a resulting solution is given (for instance, 4 gallons) but the amounts of the mixing solutions are unknown, how are the amounts of the mixing solutions represented?

Lesson Summary

- Mixture problems deal with quantities of solutions and mixtures.
- The general structure of the expressions for mixture problems are

$$\text{Whole Quantity} = \text{Part} + \text{Part}.$$

- Using this structure makes the equation resemble the following:

  $(\% \text{ of resulting quantity})(\text{amount of resulting quantity}) =$
  $(\% \text{ of } 1^{\text{st}} \text{ quantity})(\text{amount of } 1^{\text{st}} \text{ quantity}) + (\% \text{ of } 2^{\text{nd}} \text{ quantity})(\text{amount of } 2^{\text{nd}} \text{ quantity}).$

Name ____________________ Date ____________

A 25% vinegar solution is combined with triple the amount of a 45% vinegar solution and a 5% vinegar solution resulting in 20 milliliters of a 30% vinegar solution.

1. Determine an equation that models this situation, and explain what each part represents in the situation.

2. Solve the equation and find the amount of each of the solutions that were combined.

1. A 3-liter container is filled with a liquid that is 40% juice. A 5-liter container is filled with a liquid that is 60% juice. What percent of juice is obtained by putting the two mixtures together?

Let x represent the percent of juice in the resulting mixture.

$$3(0.4)+5(0.6)=8x$$

I know there will be a total of 8 liters in the resulting mixture because there is a 3-liter container and a 5-liter container.

$$1.2+3=8x$$
$$4.2=8x$$

In order to solve for x, I must collect the like terms.

$$\left(\frac{1}{8}\right)(4.2)=\left(\frac{1}{8}\right)8x$$
$$0.525=x$$

I multiply by the multiplicative inverse of 8 in order to isolate the variable.

Therefore, the resulting mixture will have 52.5% ***juice.***

2. Solution A contains 50 liters of a solution that is 75% hydrochloric acid. How many liters of Solution B containing 50% hydrochloric acid must be added to get a solution that is 60% hydrochloric acid?

Let b represent the amount of Solution B, in liters, to be added.

$$50(0.75)+(0.50)b=(0.60)(50+b)$$

The final solution will have 50 liters of Solution A and b liters of Solution B.

$$37.5+0.5b=30+0.6b$$
$$37.5+0.5b-30=30+0.6b-30$$
$$7.5+0.5b=0.6b$$
$$7.5+0.5b-0.5b=0.6b-0.5b$$
$$7.5=0.1b$$

I know that I need to collect like terms, even when they are on opposite sides of the equal sign.

$$\left(\frac{1}{0.1}\right)(7.5)=\left(\frac{1}{0.1}\right)(0.1b)$$
$$75=b$$

In order to get the desired solution, 75 ***liters of Solution B needs to be added.***

3. Miguel is consolidating two containers of trail mix. The first container has 2.5 cups and is 70% nuts. The second container is 3 cups. If the resulting trail mix is 50% nuts, what percent of the second container is nuts?

I am trying to determine the percent of nuts in the second container of trail mix.

Let n represent the percent of nuts in the second container of trail mix.

$$2.5(0.7) + 3n = 5.5(0.5)$$
$$1.75 + 3n = 2.75$$
$$1.75 + 3n - 1.75 = 2.75 - 1.75$$
$$3n = 1$$
$$\left(\frac{1}{3}\right)3n = \left(\frac{1}{3}\right)(1)$$
$$n = \frac{1}{3}$$

After consolidating, Miguel will have 5.5 cups of trail mix.

I use my algebraic knowledge to isolate the variable.

The second container of trail mix had $33\frac{1}{3}\%$ nuts.

4. Veronica wants to create a 20-cup mixture of candy with 80% of the candy being chocolate by mixing two bags of candy. The first bag of candy contains 50% chocolate candy, and the second bag contains 100% chocolate candy. How much of each bag should she use?

Let x represent the amount, in cups, in the first bag of candy.

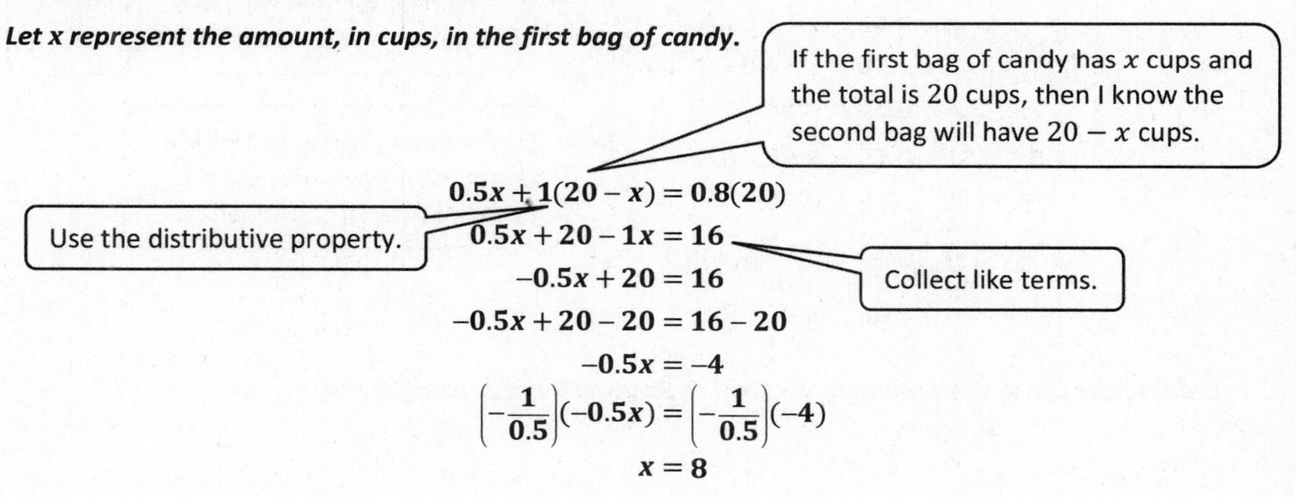

$$0.5x + 1(20 - x) = 0.8(20)$$
$$0.5x + 20 - 1x = 16$$
$$-0.5x + 20 = 16$$
$$-0.5x + 20 - 20 = 16 - 20$$
$$-0.5x = -4$$
$$\left(-\frac{1}{0.5}\right)(-0.5x) = \left(-\frac{1}{0.5}\right)(-4)$$
$$x = 8$$

Therefore, Veronica needs to use 8 cups of candy from the first bag and 12 cups of candy from the second bag to get her desired mixture.

1. A 5-liter cleaning solution contains 30% bleach. A 3-liter cleaning solution contains 50% bleach. What percent of bleach is obtained by putting the two mixtures together?

2. A container is filled with 100 grams of bird feed that is 80% seed. How many grams of bird feed containing 5% seed must be added to get bird feed that is 40% seed?

3. A container is filled with 100 grams of bird feed that is 80% seed. Tom and Sally want to mix the 100 grams with bird feed that is 5% seed to get a mixture that is 40% seed. Tom wants to add 114 grams of the 5% seed, and Sally wants to add 115 grams of the 5% seed mix. What will be the percent of seed if Tom adds 114 grams? What will be the percent of seed if Sally adds 115 grams? How much do you think should be added to get 40% seed?

4. Jeanie likes mixing leftover salad dressings together to make new dressings. She combined 0.55 L of a 90% vinegar salad dressing with 0.45 L of another dressing to make 1 L of salad dressing that is 60% vinegar. What percent of the second salad dressing was vinegar?

5. Anna wants to make 30 mL of a 60% salt solution by mixing together a 72% salt solution and a 54% salt solution. How much of each solution must she use?

6. A mixed bag of candy is 25% chocolate bars and 75% other filler candy. Of the chocolate bars, 50% of them contains caramel. Of the other filler candy, 10% of them contain caramel. What percent of candy contains caramel?

7. A local fish market receives the daily catch of two local fishermen. The first fisherman's catch was 84% fish while the rest was other non-fish items. The second fisherman's catch was 76% fish while the rest was other non-fish items. If the fish market receives 75% of its catch from the first fisherman and 25% from the second, what was the percent of other non-fish items the local fish market bought from the fishermen altogether?

Opening Exercise

You are about to switch out your books from your locker during passing period but forget the order of your locker combination. You know that there are the numbers 3, 16, and 21 in some order. What is the percent of locker combinations that start with 3?

Locker Combination Possibilities:

$3, 16, 21$

$21, 16, 3$

$16, 21, 3$

$21, 3, 16$

$16, 3, 21$

$3, 21, 16$

Example 1

All of the 3-letter passwords that can be formed using the letters A and B are as follows: AAA, AAB, ABA, ABB, BAA, BAB, BBA, BBB.

a. What percent of passwords contain at least two B's?

b. What percent of passwords contain no A's?

Exercises 1–2

1. How many 4-letter passwords can be formed using the letters A and B?

2. What percent of the 4-letter passwords contain

   a. No A's?

   b. Exactly one A?

   c. Exactly two A's?

   d. Exactly three A's?

   e. Four A's?

   f. The same number of A's and B's?

Example 2

In a set of 3-letter passwords, 40% of the passwords contain the letter B and two of another letter. Which of the two sets below meets the criteria? Explain how you arrived at your answer.

Set 1

BBB	AAA	CAC
CBC	ABA	CCC
BBC	CCB	CAB
AAB	AAC	BAA
ACB	BAC	BCC

Set 2

CEB	BBB
EBE	CCC
CCC	EEE
EEB	CBC
CCB	ECE

Exercises 3–4

3. Shana read the following problem:

   "How many letter arrangements can be formed from the word *triangle* that have two vowels and two consonants (order does not matter)?"

   She answered that there are 30 letter arrangements.

   Twenty percent of the letter arrangements that began with a vowel actually had an English definition. How many letter arrangements that begin with a vowel have an English definition?

4. Using three different keys on a piano, a songwriter makes the beginning of his melody with three notes, C, E, and G: CCE, EEE, EGC, GCE, CEG, GEE, CGE, GGE, EGG, EGE, GCG, EEC, ECC, ECG, GGG, GEC, CCG, CEE, CCC, GEG, CGC.

   a. From the list above, what is the percent of melodies with all three notes that are different?

   b. From the list above, what is the percent of melodies that have three of the same notes?

Example 3

Look at the 36 points on the coordinate plane with whole number coordinates between 1 and 6, inclusive.

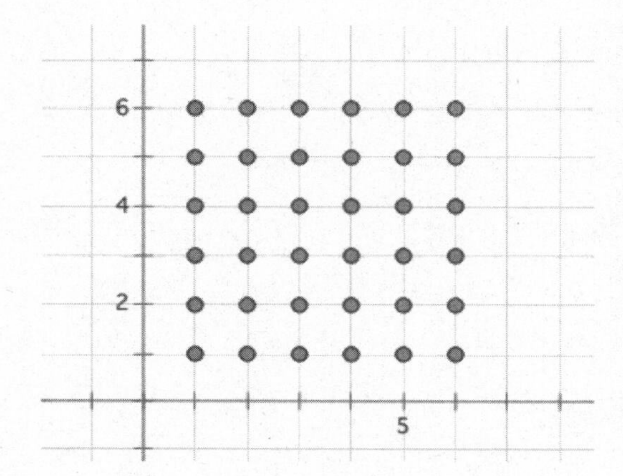

a. Draw a line through each of the points which have an x-coordinate and y-coordinate sum of 7.
   Draw a line through each of the points which have an x-coordinate and y-coordinate sum of 6.
   Draw a line through each of the points which have an x-coordinate and y-coordinate sum of 5.
   Draw a line through each of the points which have an x-coordinate and y-coordinate sum of 4.
   Draw a line through each of the points which have an x-coordinate and y-coordinate sum of 3.
   Draw a line through each of the points which have an x-coordinate and y-coordinate sum of 2.
   Draw a line through each of the points which have an x-coordinate and y-coordinate sum of 8.
   Draw a line through each of the points which have an x-coordinate and y-coordinate sum of 9.
   Draw a line through each of the points which have an x-coordinate and y-coordinate sum of 10.
   Draw a line through each of the points which have an x-coordinate and y-coordinate sum of 11.
   Draw a line through each of the points which have an x-coordinate and y-coordinate sum of 12.

b. What percent of the 36 points have coordinate sum 7?

c. Write a numerical expression that could be used to determine the percent of the 36 points that have a coordinate sum of 7.

d. What percent of the 36 points have coordinate sum 5 or less?

e. What percent of the 36 points have coordinate sum 4 or 10?

Lesson Summary

To find the percent of possible outcomes for a counting problem you need to determine the total number of possible outcomes and the different favorable outcomes. The representation

$$\text{Quantity} = \text{Percent} \times \text{Whole}$$

can be used where the quantity is the number of different favorable outcomes and the whole is the total number of possible outcomes.

Name ______________________________ Date ______________

There are a van and a bus transporting students on a student camping trip. Arriving at the site, there are 3 parking spots. Let v represent the van and b represent the bus. The chart shows the different ways the vehicles can park.

	Parking Space 1	Parking Space2	Parking Space 3
Option 1	V	B	
Option 2	V		B
Option 3	B	V	
Option 4	B		V
Option 5		V	B
Option 6		B	V

a. In what percent of the arrangements are the vehicles separated by an empty parking space?

b. In what percent of the arrangements are the vehicles parked next to each other?

c. In what percent of the arrangements does the left or right parking space remain vacant?

1. A carnival game requires you to roll a six-sided number cube two times. To determine if you win a prize, you must calculate the product of the two rolls. The possible products are

1	2	3	4	5	6
2	4	6	8	10	12
3	6	9	12	15	18
4	8	12	16	20	24
5	10	15	20	25	30
6	12	18	24	30	36

a. What is the percent that the product will be greater than 20?

Greater than 20 does not include 20.

$\frac{6}{36} = \frac{1}{6} = 16\frac{2}{3}\%$

I know the numerator represents the number of outcomes with a product greater than 20, and the denominator represents the total number of outcomes.

b. In order to win the carnival game, the product can be no more than 10. What percent chance do you have of winning the game?

No more than 10 means the product must be 10 or less.

$\frac{19}{36} = 52\frac{7}{9}\%$

I have about a 53% ***chance of winning the carnival game.***

2. Calleigh loves to accessorize. She always wears three pieces of jewelry using combinations of rings, necklaces, and bracelets. The table shows the different combinations of accessories Calleigh wore with her last eight outfits.

R	B	N	R	B	N	N	B
R	N	B	B	N	R	R	B
R	R	N	B	R	B	B	B

a. What percent of Calleigh's outfits included at least one ring?

$\frac{6}{8}=\frac{3}{4}=75\%$

There are 6 outfits where Calleigh wore at least one ring and 8 total outfits.

At least 1 ring means that Calleigh wears 1 or more rings.

b. What percent of Calleigh's outfits only included one type of accessory?

$\frac{2}{8}=\frac{1}{4}=25\%$

The first and last outfit Calleigh wore only included one type of accessory.

1. A six-sided die (singular for dice) is thrown twice. The different rolls are as follows:

   1 and 1, 1 and 2, 1 and 3, 1 and 4, 1 and 5, 1 and 6,
   2 and 1, 2 and 2, 2 and 3, 2 and 4, 2 and 5, 2 and 6,
   3 and 1, 3 and 2, 3 and 3, 3 and 4, 3 and 5, 3 and 6,
   4 and 1, 4 and 2, 4 and 3, 4 and 4, 4 and 5, 4 and 6,
   5 and 1, 5 and 2, 5 and 3, 5 and 4, 5 and 5, 5 and 6,
   6 and 1, 6 and 2, 6 and 3, 6 and 4, 6 and 5, 6 and 6.

   a. What is the percent that both throws will be even numbers?
   b. What is the percent that the second throw is a 5?
   c. What is the percent that the first throw is lower than a 6?

2. You have the ability to choose three of your own classes, art, language, and physical education. There are three art classes (A1, A2, A3), two language classes (L1, L2), and two P.E. classes (P1, P2) to choose from. The order does not matter and you must choose one from each subject.

A1, L1, P1	A2, L1, P1	A3, L1, P1
A1, L1, P2	A2, L1, P2	A3, L1, P2
A1, L2, P1	A2, L2, P1	A3, L2, P1
A1, L2, P2	A2, L2, P2	A3, L2, P2

   Compare the percent of possibilities with A1 in your schedule to the percent of possibilities with L1 in your schedule.

3. Fridays are selected to show your school pride. The colors of your school are orange, blue, and white, and you can show your spirit by wearing a top, a bottom, and an accessory with the colors of your school. During lunch, 11 students are chosen to play for a prize on stage. The table charts what the students wore.

Top	W	O	W	O	B	W	B	B	W	W	W
Bottom	B	O	B	B	O	B	B	B	O	W	B
Accessory	W	O	B	W	B	O	B	W	O	O	O

   a. What is the percent of outfits that are one color?
   b. What is the percent of outfits that include orange accessories?

4. Shana wears two rings (G represents gold, and S represents silver) at all times on her hand. She likes fiddling with them and places them on different fingers (pinky, ring, middle, index) when she gets restless. The chart is tracking the movement of her rings.

	Pinky Finger	Ring Finger	Middle Finger	Index Finger
Position 1		G	S	
Position 2			S	G
Position 3	G		S	
Position 4				S,G
Position 5	S	G		
Position 6	G	S		
Position 7	S		G	
Position 8	G		S	
Position 9		S,G		
Position 10		G	S	
Position 11			G	S
Position 12		S		G
Position 13	S,G			
Position 14			S,G	

a. What percent of the positions shows the gold ring on her pinky finger?

b. What percent of the positions shows both rings on the same finger?

5. Use the coordinate plane below to answer the following questions.

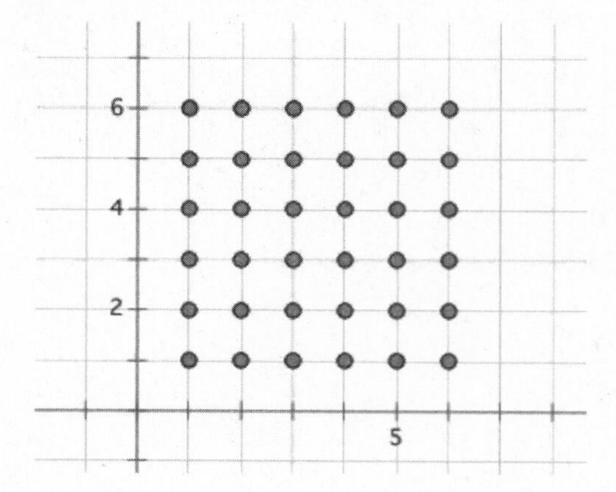

a. What is the percent of the 36 points whose quotient of $\frac{x\text{-coordinate}}{y\text{-coordinate}}$ is greater than one?

b. What is the percent of the 36 points whose coordinate quotient is equal to one?

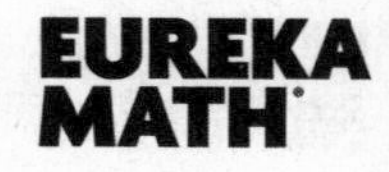

Credits

Great Minds® has made every effort to obtain permission for the reprinting of all copyrighted material. If any owner of copyrighted material is not acknowledged herein, please contact Great Minds for proper acknowledgment in all future editions and reprints of this module.